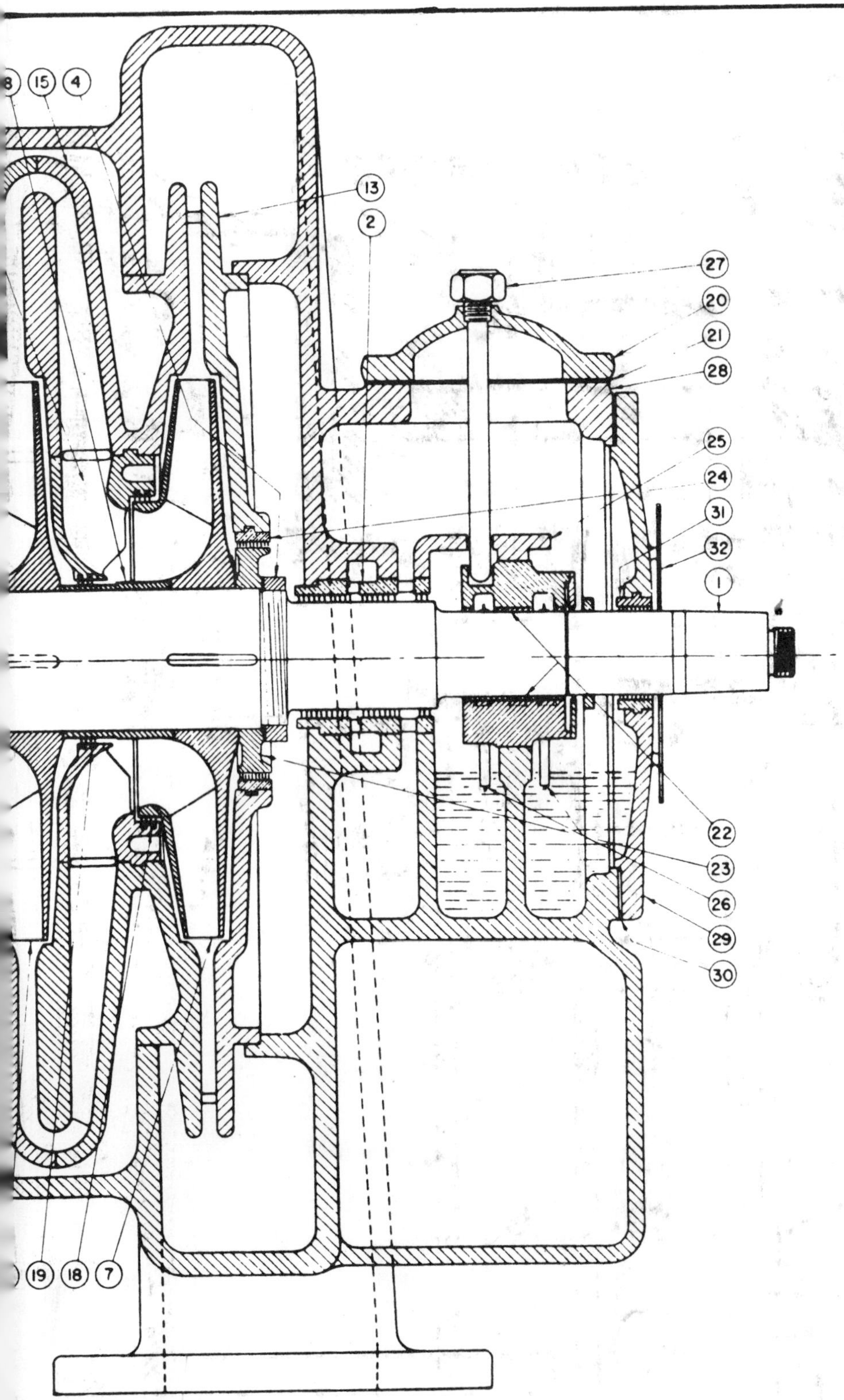

COMPRES

1. SHAFT
2. SHAFT LABYRINTH – EITHER END
3. IMPELLER LOCK NUT – SUCTION END
4. IMPELLER LOCK NUT – DISCHARGE END
5. IMPELLER – FIRST STAGE
6. IMPELLER – SECOND STAGE
7. IMPELLER – THIRD STAGE
8. IMPELLER SPACERS
9. INLET GUIDE VANE – FIRST STAGE
10. INLET GUIDE VANE – SECOND STAGE
11. INLET GUIDE VANE – THIRD STAGE
12. INTAKE WALLS
13. DISCHARGE WALL
14. DIAPHRAGM – FIRST STAGE
15. DIAPHRAGM – SECOND STAGE
16. INLET LABYRINTH – FIRST STAGE
17. INLET LABYRINTH – SECOND STAGE
18. INLET LABYRINTH – THIRD STAGE
19. SPACER LABYRINTHS

SEAL END

20. BEARING INSPECTION COVER
21. BEARING INSPECTION COVER GASKET
22. BEARING ASSEM. (UPPER & LOWER HALF)
23. BALANCING PISTON
24. BALANCING PISTON LABYRINTH
25. BEARING CAP
26. OIL RINGS (2)
27. THERMOMETER WELL
28. SEAL HOUSING
29. SEAL HOUSING COVER
30. SEAL HOUSING COVER GASKET
31. SEAL LABYRINTH
32. SHAFT END BAFFLE

THRUST END

34. BEARING INSPECTION COVER
35. BEARING INSPECTION COVER GASKET
36. BEARING ASSEM. (UPPER & LOWER HALF)
37.
38.
39. BEARING CAP
40. OIL RINGS (2)
41. THERMOMETER WELL
42. OIL CHAMBER COVER
43. OIL CHAMBER COVER GASKET
44. OIL PUMP ASSEMBLY
45. OIL DISCHARGE PIPE
46. THRUST END COVER
47. THRUST DISC
48. KINGSBURY THRUST BEARING
49. THRUST BEARING HOUSING
50. OIL STRAINER
51. OIL STRAINER CAP
52. PUMP DRIVE OIL NOZZLE
53. WORM LOCK NUT
54. WORM SPACER
55. SHAFT
56. BRACKET
57. WORM GEAR
58. BUSHING
59. SHAFT LOCK NUT
60. WORM GEAR LOCK NUT
61. WORM
62. RELIEF VALVE
63. SIGHT GLASS

Carrier
SYRACUSE, N.Y.

TITLE CARRIER CENTRIFUGAL COMPRESSOR (AIR)

DR. J C H	1-7-43	SCALE
CK.		NONE
DWG. No.	18P64-124	

(Courtesy of Elliott Company)

COMPRESSOR PERFORMANCE

COMPRESSOR PERFORMANCE

Selection, Operation, and Testing
of Axial and Centrifugal Compressors

M. THEODORE GRESH, P.E.

Butterworth–Heinemann
Boston London Singapore Sydney Toronto Wellington

Recognizing the importance of preserving what has been written, it is the policy of Butterworth–Heinemann to have the books it publishes printed on acid-free paper, and we exert our best efforts to that end.

Library of Congress Cataloging-in-Publication Data

Gresh, M. Theodore.
Compressor performance : selection, operation, and testing of axial and centrifugal compressors / M. Theodore Gresh.
p. cm.
Includes bibliographical references.
ISBN 0-409-90237-3
1. Compressors—Aerodynamics. 2. Compressors—Testing.
I. Title.
TJ267.5.C5G74 1991
621.5'1—dc20 90-1804
CIP

British Library Cataloguing in Publication Data

Gresh, M. Theodore
Compressor performance : selection, operation, and testing of axial and centrifugal compressors.
1. Compressors
I. Title
621.51

ISBN 0-409-90237-3

Butterworth–Heinemann
80 Montvale Avenue
Stoneham, MA 02180

10 9 8 7 6 5 4 3 2 1

Printed in the United States of America

To Kathy
Thanks

Contents

PART III REFERENCE MATERIAL

Preface

This text has been designed to be used primarily by equipment users, as a guide in selecting, monitoring, and enhancing the aerodynamic performance of various types of compressors. Some basic theory is included as an aid in helping field personnel to better understand the aerodynamics of compressors so that performance enhancements and trouble resolution can be more readily realized. As much as possible, I have attempted to stick to the "business end" of the applicable aerodynamic principles.

This book is the result of various books, articles, notes, seminars, and personal experience that I have collected over the years working in the field of compressor aerodynamics. As it is such a "collection," references have been used extensively as noted.

The concepts and procedures presented in the following pages, while generally in line with Elliott Company Policy and Industry Standards, include opinions belonging solely to me. Conforming to guidelines in this text therefore does not mean compliance with Elliott Company, API, or other industry standards. The methods presented are meant to be guidelines used for day-to-day performance trending or as the first step in selection, trouble-shooting, or retrofitting equipment. For potential warranty cases, customer and vendor must agree on a specific test procedure before proceeding. For an "out of warranty" problem the field engineer is best advised to get some help from the equipment manufacturer, after some initial analysis is completed.

M. T. Gresh

Acknowledgments

A special thanks to Elliott Company for the support in this endeavor and permission to print the various drawings and articles used throughout this book. Thanks also to GHH, Sultzer, Carrier, General Electric Co., Ebara, Demag, Ingersol Rand, DuPont, Gulf Publishing, and Natural Gas Processors Suppliers Association for technical data provided.

Thanks also to all the people who have personally helped by offering their technical support and encouraging comments, with special thanks to Marc Sassos, Dave Galster, Don Rudisel, Ron Aungier, John Beaty, Bob Spigarelli, Paul Gallick, Bill Hohlweg, Katsuhiko Yamanaka, Dennis Maffessanti, Ross Hackel, and my wife, Lynn.

The talents of Frank Weidler and Tom Humphrey are displayed throughout this book in the various drawings they created for me.

Most crucial, though, to the development of this book has been the excellent cooperation and typng skills of Kathy Lazur.

Symbols

A	Area, ft^2
a	Speed of sound, ft/sec
BHP	Brake or shaft horsepower
C	Discharge coefficient
c_p	Specific heat at constant pressure BTU/lb mole °R
c_v	Specific heat at constant volume BTU/lb mole °R
D	Pipe diameter, inches
d	Throat, or impeller diameter, inches
E	Voltage
E	Velocity of approach factor
Eff	Efficiency
Fa	Orifice meter thermal expansion factor
g_c	Gravitational constant $32.2 \frac{\text{ft-lb mass}}{\text{lb force-sec}^2}$
GHP	Gas horsepower
H	Head $\frac{\text{ft-lbs force}}{\text{lb mass}}$
HP	Horsepower
h	Enthalpy (BTU/lb mass)
h_w	Differential pressure, inches water
I	Amperage
K	Orifice meter flow coefficient
k	Adiabatic exponent (c_p/c_v)
MW	Molecular weight
$\dot{M}$	Weight flow (lb/min)
M	Mach number, V/a
N	Speed, RPM
N_s	Specific speed
n	Polytropic exponent
P	Static pressure (psia)
P_c	Critical pressure (psia)
P_r	Reduced pressure
P_T	Total pressure, psia
P_0	Stagnation pressure, psia
P_v	Velocity pressure
PF	Power factor
Q	Flow rate ft^3/min
Q_s	Flow rate ft^3/sec
q	Heat transfer ft-lb force/lb mass
R	Gas constant (1544/MW)
Re	Reynolds number
r_p	Pressure ratio (P_2/P_1)
S	Entropy
SHP	Shaft horsepower
T	Absolute temperature (°Rankine = °F + 459.6)
T_c	Critical temperature (°Rankine)
T_R	Reduced temperature (T/T_c)
t	Temperature (°F)
U	Tip speed, FPS
u	Internal energy, ft-lb force/lb mass
V	Velocity (ft/sec)
v	Specific volume (ft^3/lb mass)
W	Work $\frac{\text{ft-lbs force}}{\text{lb mass}}$
Y	Orifice meter expansion factor
Ya	Adiabatic expansion factor
Z	Compressibility factor
z	Verticle height

GREEK LETTERS

β	Throat (or orifice) to pipe diameter ratio
η	Efficiency
γ	Work coefficient
μ	Head coefficient
μ'	Absolute viscosity lb-sec/ft^2
ν'	Kinematic viscosity ft^2/sec
ρ	Density lb/ft^3
ϕ	Flow coefficient

SUBSCRIPTS

ad	Adiabatic process (H_{ad})
p	Polytropic process (H_p)
S	Standard conditions—usually 14.7 psia, 60°F, dry air
1	Inlet conditions (P_1)(Q_1)(t_1)
2	Discharge conditions (T_2)(P_z)

I

THEORY

1

INTRODUCTION TO AERODYNAMICS

Down through the years, human needs and desires have required a continued evolution of more and more sophisticated fluid-handling apparatus. In general, fluid handling involves two problems, fluid transportation and fluid pressurization.

Ancient man was most concerned with liquid transport and storage. Of primary concern was irrigation for agricultural purposes and transport of water to cities.

The Bronze Age, which began about 3000 B.C., brought with it the requirement of mechanisms for enhancing air supply to hearth furnaces. Air was first introduced in hearths by crude drafts and simple fanning. With time, innovation brought improved air supply devices. Hearths were oriented to capture the prevailing winds, and chimneys were added to help draw more air to the furnaces.

With the advent of the Iron Age, which began around 1000 B.C., no longer were simple drafting techniques adequate. A much higher hearth temperature required a pressurized air blast. Small foot- and hand-operated bellows were used in the small hearths of the farrier and blacksmith. Five hundred years ago immense bellows were used in Germany to supply the air required for large furnaces. These were ultimately supplemented by piston pumps. Today, rotary compressors are used almost exclusively for this purpose.

The Industrial Revolution and, most recently, the Space Age, have produced an exponential growth in the advancement of turbomachinery, from the simple squirrel cage fan in a car's heater to the liquid fuel pumps used on the space shuttle engines.

FLUID MECHANICS AND THERMODYNAMICS

Little heed was paid to the various fluid properties in the design of compression devices until the 19th century. Until this period, only a slight density and temperature change was encountered at the reduced compression ratios used in air pumps. The designer had a large margin of error possible since he was at liberty to "tinker" and adjust the apparatus at the job site until it was perfected. In most instances both the building and design were done at the job site.

Concepts of flow, energy, work, heat, and momentum, which eluded the grasp of the early Greek philosophers and later the Roman engineers, gradually began to be understood and interpreted under the impetus of the Renaissance scientists da Vinci, Galileo, Newton, Bernoulli, Euler, St. Venant, Stokes, and Navier. The mathematical tools to describe and solve problems were wrought by Liebniz, Newton, De Moivre, Descartes, Legendre, and others. Watt, Stephenson, Carnot, Clausins, and Thurston through their applied efforts on the steam locomotive developed technical, mechanical, and thermodynamic solutions which have contributed to the compression equipment of our century. The science of heat transfer, thermodynamics, and energy conservation was developed by Maxwell, Thurston, Otto, Helmholtz, Steffan, Boltzmann, Rayleigh, Rankine, Mach, and Plank. In the wake of the Wright Brothers' first flight at Kitty Hawk came the aerodynamic scientists Kutta, Joukowsky, Von Karman, Von Mises, Prandtl, Lamb, Struhal, Tiejens, Stodola, Dryden, Parsons, and Paulson. With the advent of flight, these men developed theories on boundary layer, vortex shedding, aeroelastic phenomena, and other necessary tools used in the design of present-day turbomachinery [1].

FIRSTS

In Alexandria, Egypt, about 130 A.D., a priest scientist named Hero employed aerothermo principles to generate steam and drive a small reaction turbine.

Although the fluid mechanics of a compressor and turbine are much the same, knowledge of fluid mechanics is much more crucial for the design of a compressor than for a turbine. A turbine, with its flow usually going from a high to a low pressure, will always work. With reasonable design, it will work at a respectable efficiency. A compressor, conversely, particularly an axial compressor, will not produce any pressure rise at all unless properly designed. Consequently, very little activity was seen in the field of compressor design until the 18th century.

In 1705 Denis Papin published full descriptions of the centrifugal blowers and pumps he had developed; however, the efficiency of these machines is unknown [2,3].

John Barber designed and patented a gas turbine engine in England in 1791. The engine was designed to operate on a constant pressure cycle using gas from wood or coal as fuel [4].

In 1851, Henry Gifford flew from Paris to Trappes in the first successful aircraft propulsion device, a propeller-driven dirigible balloon powered by a steam engine [5].

In 1872, Dr. Stolze patented a gas turbine which was eventually built and operated. The engine employed a multi-stage axial-flow compressor and a multi-stage turbine with both mounted on the same shaft. Heat was supplied to the air by means of a furnace located between the compressor and turbine [4].

Around the same period, Parsons and Delaval developed a reaction steam turbine, for the purpose of driving blowers and generators. Although Parsons also used this device in reverse to serve as a compressor, the efficiency was low—around 60%. Sir Charles Parsons' 1884 patent also made reference to the gas turbine engine and provided for cooling to the turbine blades [1–3].

The first United States patent covering a gas turbine was by Charles Curtis (inventor of the Curtis steam turbine) in June of 1895 [3].

In 1905, Dr. Alfred J. Buchi of Switzerland first suggested the turbocharger for enhancing the output of internal combustion engines. He later went on to patent his ideas in 1915 and to organize the Buchi Syndicate in 1927 for the purpose of developing his systems [3].

It was not until January 16, 1930, that Frank Wittle, an officer in Great Britain's Royal Airforce, developed and patented a practical design for an aircraft gas turbine engine. However, the British Air Ministry dismissed the design, finding it impractical [3,6].

A few years later in 1934, a German named Hans von Ohain began development of an engine of similar design. In 1936 he joined forces with Ernst Heinkel, an airplane manufacturer. Progress was good and an aircraft with von Ohain's engine was successfully flown in August, 1939. Von Ohain's HES8A Engine had a centrifugal compressor and a mixed-flow expander [2,6].

Meanwhile, Wittle had obtained some money from the British Air Ministry to develop his engine. In May 1941, an aircraft with Wittle's jet engine was successfully flown. Wittle's W2/700 Turbojet Engine, which consisted of an axial compressor, a single-stage centrifugal compressor, and an axial expander, was eventually developed into the Rolls Royce Welland in England and also the General Electric J33 in the United States [3,6].

DEFINITION OF COMPRESSOR

A compressor is a device that transfers energy to a gaseous fluid for the purpose of raising the pressure of the fluid as in the case where the compressor is the prime mover of the fluid through the process. The purpose may also include a desired temperature rise to enhance the chemical reaction in the process.

Devices that develop less than 5.0 psig, or that effect a 7% density increase from inlet to discharge, are classified as fans or blowers. Above this level, the devices are referred to as compressors. Due to the low density change, fan equations assume constant density, thus simplifying the calculations [7,8].

Pumps are very similar to compressors but deal primarily with incompressible hydraulic fluids, whereas compressors generally deal with compressible gaseous fluids.

TYPES OF COMPRESSORS

The two basic types of compressors are positive displacement and dynamic.

POSITIVE DISPLACEMENT COMPRESSOR

The positive displacement compressor functions by means of entrapping a volume of gas and reducing that volume, as in the common bicycle pump, and the screw compressor shown in Figure 1.1. The general characteristics of the positive displacement compressor are constant flow and variable pressure ratio (for a given speed).

Positive displacement compressors include

- piston compressor
- screw compressor
- vane compressor
- lobe compressor

DYNAMIC COMPRESSOR

The dynamic compressor depends on motion to transfer energy from the compressor rotor to the process gas. The characteristics of compression vary depending on the type of dynamic compressor and on the type of gas being com-

Figure 1.1. Positive displacement compressor. *(Courtesy of MAN GHH.)*

pressed. The flow is continuous. There are no valves and there is no "containment" of the gas, as in a positive displacement compressor. Compression depends on the dynamic interaction between the mechanism and the gas.

Dynamic compressors include

- ejector
- centrifugal compressor
- axial compressor

Ejector An ejector is a very simple device which uses a high-pressure jet stream to compress gas. The momentum of the high-pressure jet stream is transferred to the low-pressure process gas. This type of compressor is commonly used for vacuum applications.

Centrifugal Compressor A centrifugal compressor acts on a gas by means of blades on a rotating impeller. The rotary motion of the gas results in an outward velocity due to centrifugal forces. The tangential component of this outward velocity is then transformed to pressure by means of a diffuser.

Figure 1.2 is typical of a single-stage centrifugal compressor. A high-pressure multi-stage compressor is shown in Figure 1.3.

Axial Compressor An axial compressor imparts momentum to a gas by means of a cascade of airfoils. The lift and drag coefficients of the airfoil shape determine the compressor characteristics. Figure 1.4 shows a typical axial compressor. An axial compressor incorporated in a turbocharger is shown in Figure 1.5.

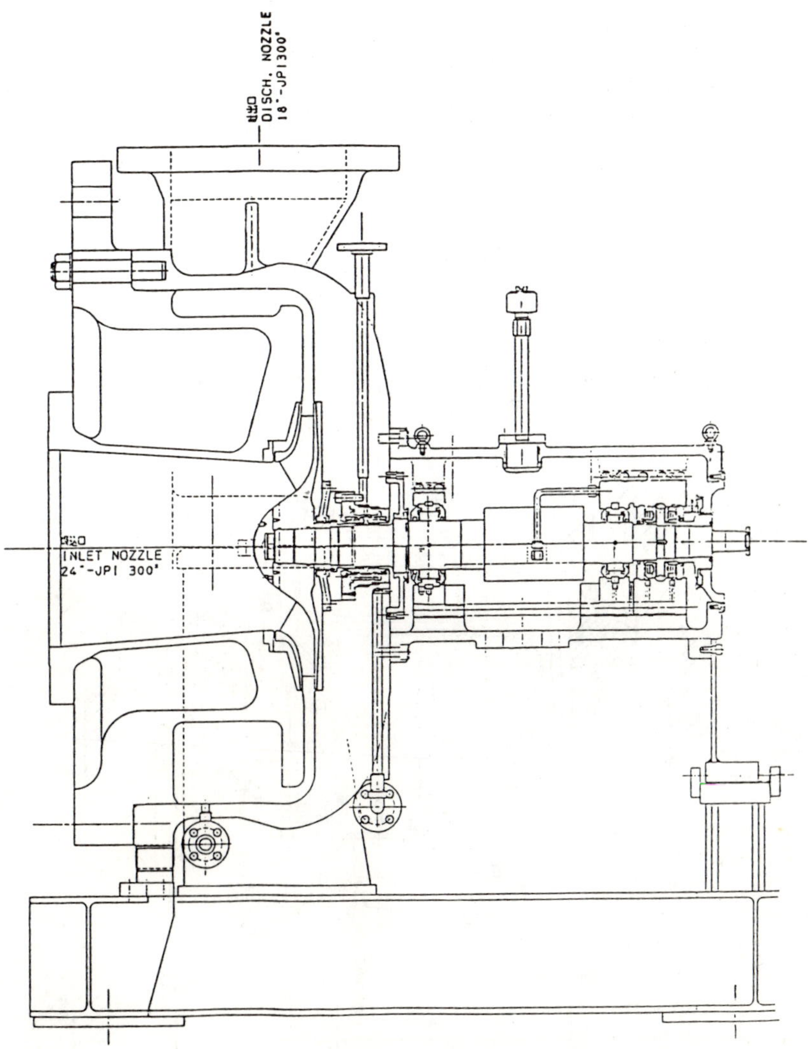

Figure 1.2. Centrifugal compressor. *(Courtesy of Ebara Corporation.)*

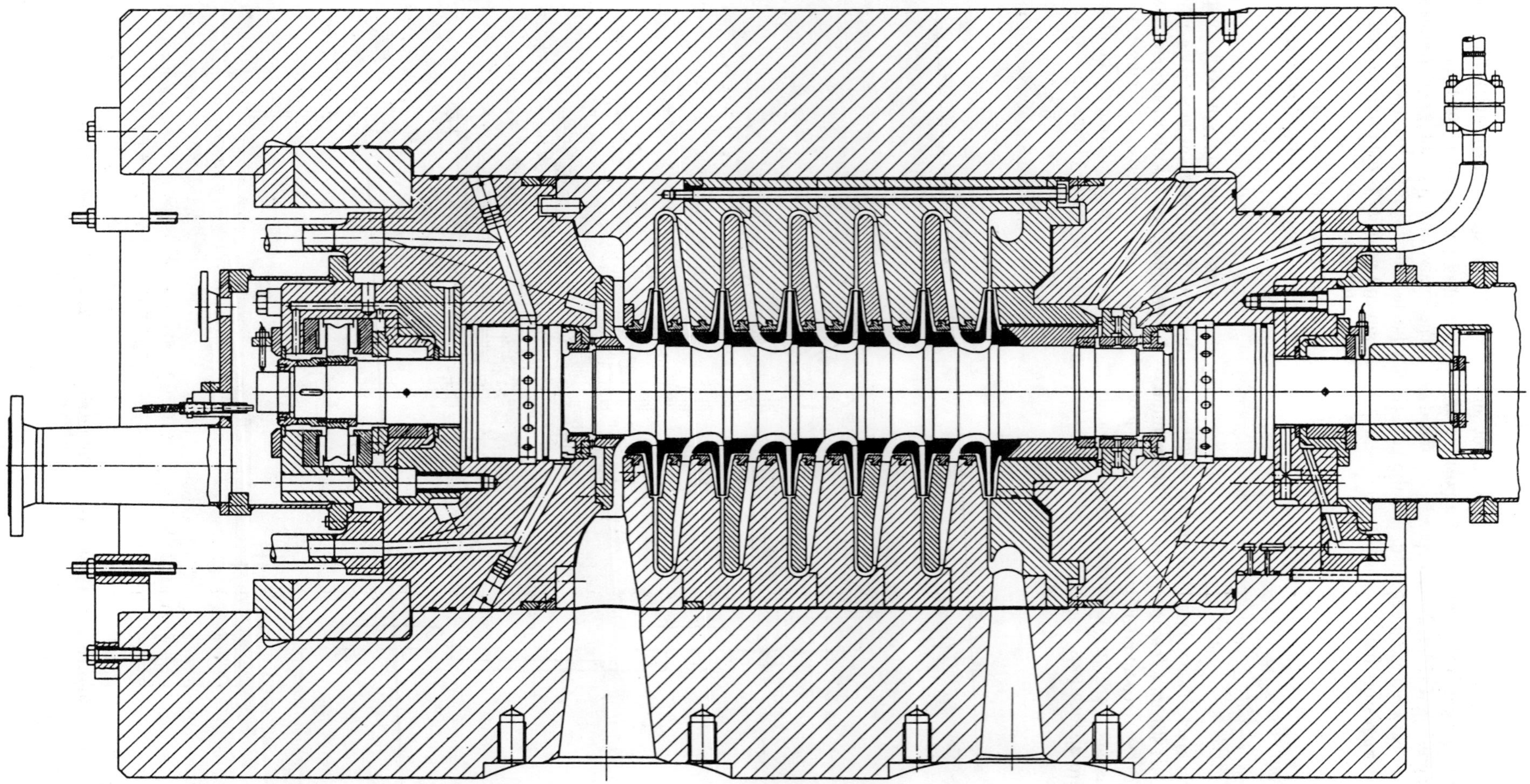

Figure 1.3. High-pressure barrel type multi-stage compressor. *(Courtesy of Mannesmann Demag.)*

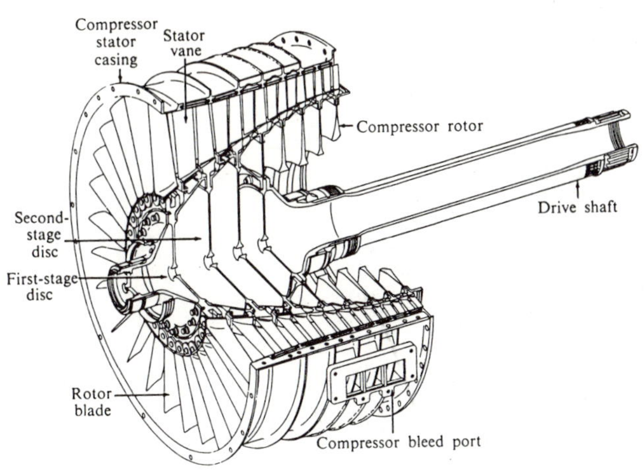

Figure 1.4. Axial compressor. *(Courtesy of General Electric Company.)*

RELATIVE COMPARISONS OF VARIOUS COMPRESSOR TYPES

CAPACITY

Axial compressors have the greatest capacity for a given volumetric size. The design is a very compact, light-weight compressor which can handle large volumes of gas. This explains its popularity for use on aircraft.

EFFICIENCY

Figure 1.6 illustrates the relative nominal efficiencies for the various types of compressors.

For small capacities, the positive displacement compressor is generally the best. At higher capacities, valve and seal leakage, mechanical friction, and flow discontinuities increase rapidly, limiting overall efficiency.

In a centrifugal compressor the opposite is true. In small capacities, the sealing surface is large in comparison to the compression element, the impeller. As the compressor size increases, the seal leakage rate grows slowly relative to volume through-put. Reduced mechanisms (bearings, valves, seals) and improved through-flow contribute to improved efficiencies at the high capacities.

Axial compressors have the best efficiency. Both mechanical and aerodynamic losses for an axial compressor are very low, resulting in efficiencies approaching 90% or even better.

Due to the configuration of the axial, the sealing surface is very small in comparison to the volume of gas flow. Also, the "wetted perimeter" (frictional surface versus the volume flow) is very small, contributing to low losses and high efficiency for large capacities. Further improvement is via the constant nominal through-velocity. Losses due to acceleration and deceleration are limited.

The boundaries shown on Figure 1.6 are constantly being changed by enhancements to compressor design such as abradable seals to reduce leakage, low friction bearings and seals, and hybrid compressor elements.

PRESSURE RATIO

Positive displacement compressors and ejectors can have a very high pressure ratio. For dynamic compressors, the centrifugal compressor achieves the highest per stage pressure ratio. Axial compressors develop very low pressure ratio per stage, thus the need for many stages.

OPERATION

Specialized training is required to operate both centrifugal and axial compressors. The primary concern is to avoid operation in aerodynamically unstable regions, including surge and rotating stall. Operation in these areas can cause equipment damage.

CHARACTERISTIC CURVES

Figure 1.7 shows the normal characteristic curve shapes for the various compressor types. Positive displacement compressors exhibit a constant volume and a variable pressure ratio, while low-speed centrifugal compressors approach being constant pressure ratio and variable volume machines. Axial compressor characteristics are somewhere in between. Very high-speed centrifugal compressors may have characteristics approaching those of axials.

An in-depth understanding of aerodynamics is required to properly design, select, operate, and maintain centrifugal and axial compressors. The advantages and disadvantages of each type are listed in Table 1.1. This table, along with the following discussions, will provide an overview of aerodynamics for these compressors. The goal is to provide the user with a better understanding of how to more efficiently operate and maintain dynamic compressors. These same tools will be useful in troubleshooting, rerating, and selecting new equipment.

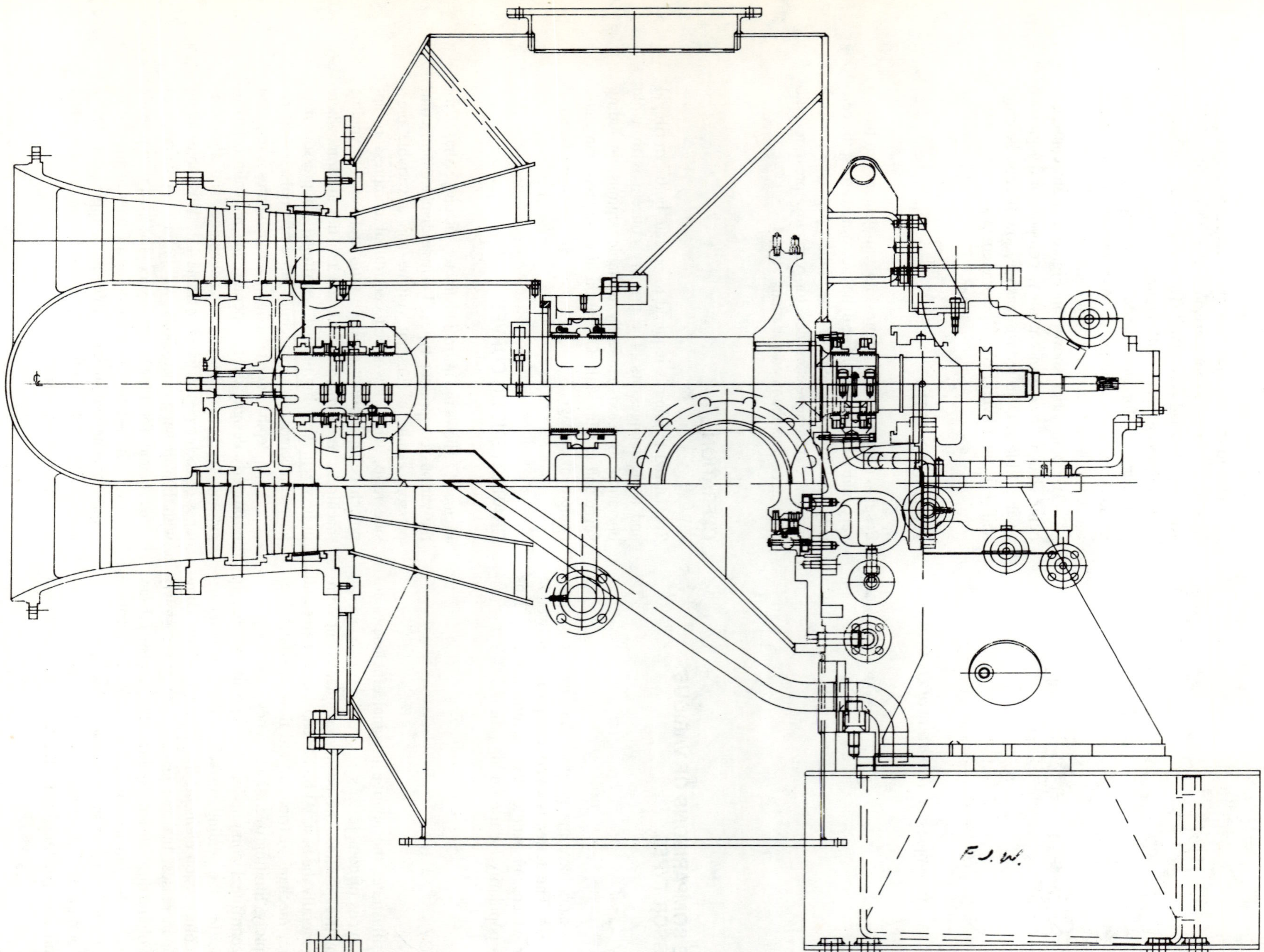

Figure 1.5. Turbocharger with axial compressor. This unit is used to supply air to a steam boiler unit. *(Reprinted with permission of Elliott Company, Jeannette, PA.)*

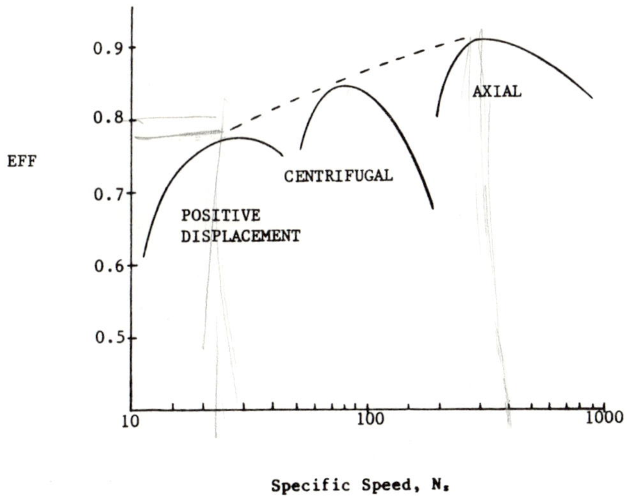

Figure 1.6. Efficiency versus type of compressor. *(Adapted from [9].)*

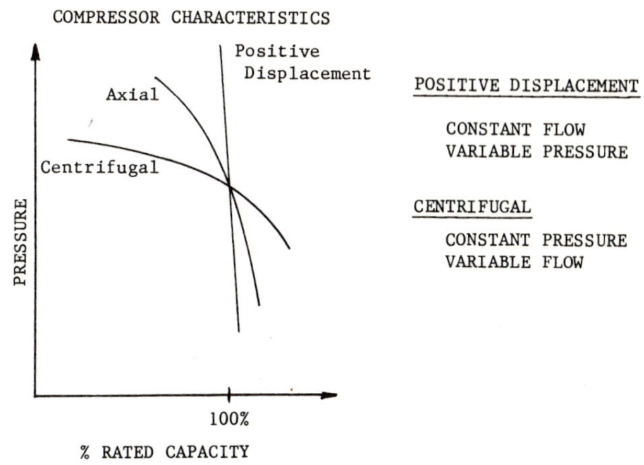

Figure 1.7. Characteristic curves. *(Data from [10].)*

TABLE 1.1 Relative Comparison of Compressors

Type	Advantages	Disadvantages
Centrifugal	Wide operating range Low maintenance High reliability	Unstable at low flow Moderate efficiency
Axial	High efficiency High-speed capability Higher flow for given size	Low pressure ratio per stage Narrow flow range Fragile and expensive blading
Positive displacement	Pressure ratio capability not affected by gas properties Good efficiencies at low specific speed	Limited capacity High weight-to-capacity ratio
Ejector	Simple design Inexpensive No moving parts High-pressure ratio	Low efficiency Requires high-pressure source

(Adapted from [10].)

2

THERMODYNAMICS

As noted in Chapter 1, knowledge of fluid mechanics and thermodynamics is very important in the proper design of a centrifugal compressor, especially if good efficiency is desired. This knowledge is especially important for the design of an axial compressor. This chapter provides a summary of many of the basic laws and equations used in the design, selection, and operation of centrifugal and axial compressors.

GAS LAWS*

No gas conforms exactly to the ideal gas laws. However, most gases conform to these laws with sufficient accuracy to yield good engineering answers. These laws, therefore, are used to form the foundation of compressor thermodynamics.

BOYLE'S LAW

The Pv product remains constant at constant temperature.

$$P_1 v_1 = P_2 v_2, \; T = \text{constant} \tag{2.1}$$

CHARLES' LAW

The volume of a gas varies directly as the absolute temperature at constant pressure.

$$\frac{v_2}{v_1} = \frac{T_2}{T_1}, \; P = \text{constant} \tag{2.2}$$

DALTON'S LAW

In a mixture of gases, the summation of partial pressures equals the total pressure of the mixture.

AVOGADRO'S LAW

All gases have the same number of moles in the same volume at the same pressure and temperature.

$$\frac{Pv}{T} = \text{constant} \tag{2.3}$$

*Adapted from "Compressor Refresher," Elliott Company, Jeannette, PA, 1975, with permission [11].

THE IDEAL GAS LAW

From Charles' and Boyle's laws

$$v = \frac{RT}{P} \tag{2.4}$$

or

$$\rho = \frac{P}{RT} \tag{2.5}$$

where

v = specific volume, ft³/lb
ρ = density, lb/ft³
$R = R_0/\text{MW} = 1545.32/\text{MW}$, ft-lb$_f$/lb-mole°R

R_0 = Universal Gas Constant
= 1.98587 BTU/lb-mole-°R
= 1545.32 ft-lb$_f$/lb-mole-°R
= 8.3143 joules/gm-mole-°K
= 10.73 psia-ft³/lb-mole-°R
MW = Gas Molecular Weight

Most gases at low pressures act as ideal gases.

Note that for a specific gas, the ideal gas equation reduces to

$$\frac{P_1 v_1}{T_1} = \frac{P_2 v_2}{T_2} \tag{2.6}$$

COMPRESSIBILITY

Most gases encountered in industrial compression do not follow the perfect gas equation of state exactly but differ in varying degrees [11]. The degree in which any gas varies from the ideal is expressed by a factor (compressibility) which modifies $Pv = RT$ to

$$Pv = ZRT \tag{2.7}$$

or

$$Z = \frac{Pv}{RT} \tag{2.8}$$

There have been many equations and charts used to determine the value of Z for any given state point. The chart most commonly used today is the Obert-Nelson chart, which in modified form appears in the section on Gas Properties in Appendix A, Figure A.1.

In order to use the compressibility chart, reduced pressure, P_R, and reduced temperature, T_R, must be determined. These values are found by dividing the state point under consideration by the respective critical values of pressure or temperature. Values for commonly used gases are listed in Appendix A, Table A.1.

BERNOULLI'S EQUATION

For incompressible flow

$$Pv + Z + \frac{V^2}{2g} = \text{constant} \tag{2.9}$$

This equation says that the sum of the pressure energy, potential energy, and kinetic energy is constant [12]. It is useful for a system where there is no work done, no heat exchanged, and therefore no change in internal energy such as for frictionless flow through a nozzle. The terms in this equation are in units of length and are frequently called pressure, elevation, and velocity head. Between two separate points in a flow

$$P_1v + Z_1 + \frac{V_1^2}{2g} = P_2v + Z_2 + \frac{V_2^2}{2g} = \text{constant} \tag{2.10}$$

Because gases are compressible, this equation is never exactly correct, but in many cases the change in density is insignificant or can be compensated for.

For determining the total or stagnation pressure P_0, $V_2 = 0$,

$$P_1v + \frac{V_1^2}{2g} = P_2v \tag{2.11}$$

or

$$P + \frac{\rho V^2}{2g} = P_0 \tag{2.12}$$

The term $\rho V^2/2g$ is called the dynamic or velocity pressure (or head). The sum of the static and dynamic pressures is equal to the total or stagnation pressure for an ideal gas in incompressible flow.

MODIFIED BERNOULLI EQUATION

Adding friction to Bernoulli's equation, represented as head loss between points 1 and 2,

$$\left(P_1v + Z_1 + \frac{V_1^2}{2g}\right) - \left(P_2v + Z_2 + \frac{V_2^2}{2g}\right) = H_{1-2} \tag{2.13}$$

For horizontal flow at constant velocity

$$Z_1 = Z_2$$
$$V_1 = V_2$$

therefore

$$H = (P_1 - P_2)v \tag{2.14}$$

THE GENERAL ENERGY EQUATION

By conducting a complete energy balance around a system we obtain the General Energy equation:

$$q + u_1 + P_1v_1 + \frac{V_1^2}{2g} + Z_1 = W + u_2 + P_2v_2 + \frac{V_2^2}{2g} + Z_2 \tag{2.15}$$

$$h = u + Pv$$

$$g + h_1 + \frac{V_1^2}{2g} + Z_1 = W + h_2 + \frac{V_2^2}{2g} + Z_2$$

Assume: q (heat transfer) $= 0$, and negligible elevation and velocity effects.

$$W(\text{work}) = h_1 - h_2 \tag{2.16}$$

THERMODYNAMIC RELATIONS FOR A PERFECT GAS

For a perfect gas [12]

c_v = Specific heat at constant volume

$$= \frac{du}{dT}$$

c_p = Specific heat at constant pressure

$$= \frac{dh}{dt}$$

By definition

$$h = u + Pv \tag{2.17}$$

or, since $Pv = RT$

$$h = u + RT$$

Therefore:

$$c_p = \frac{d(u + RT)}{dT} \tag{2.18}$$

$$= c_v + R$$

or

$$R = c_p - c_v \tag{2.19}$$

also

$$k = \frac{c_p}{c_v} \tag{2.20}$$

Combining these equations,

$$c_p = \frac{k}{k-1}R \tag{2.21}$$

and

$$c_v = \frac{R}{k-1} \tag{2.22}$$

ADIABATIC PROCESS

An adiabatic process is defined as a process in which no heat transfer takes place. This does not mean that the temperature is constant, but rather that no heat is transferred into or out from the system. In compressor theory, the terms adiabatic (no heat transfer) and isentropic (constant entropy) are used interchangeably. This is quite valid for the context in which they are used. (The actual definition of an isentropic process is an adiabatic, reversible process.)

For an adiabatic process [12],

$$c_v dT = -P dv$$

and

$$c_p dT = v\ dp$$

Combined,

$$\frac{dP}{P} = -\frac{c_p}{c_v}\ \frac{dv}{v} = -k\frac{dv}{v}$$

For k = constant,

$$\ln P = \ln v^{-k} + \ln \text{constant}$$

$$Pv^k = \text{constant},\ k = \frac{c_p}{c_v} \tag{2.23}$$

or

$$\frac{P_1}{P_2} = \left(\frac{v_2}{v_1}\right)^k \tag{2.24}$$

Combining this with the equation of state for a perfect gas,

$$\frac{T_1}{T_2} = \left(\frac{v_2}{v_1}\right)^{k-1} \tag{2.25}$$

$$\frac{T_1}{T_2} = \left(\frac{P_1}{P_2}\right)^{(k-1)/k} \tag{2.26}$$

$$\frac{v_1}{v_2} = \left(\frac{P_2}{P_1}\right)^{1/k} \tag{2.27}$$

POLYTROPIC PROCESS

When a gas follows a reversible process that includes heat transfer, the process generally proceeds such that a plot of log P versus log v is a straight line, as shown in Figure 2.1.

From this can be written [13]

$$\frac{d\ \ln P}{d\ \ln v} = -n$$

$$d\ \ln P + n\ d\ \ln v = 0$$

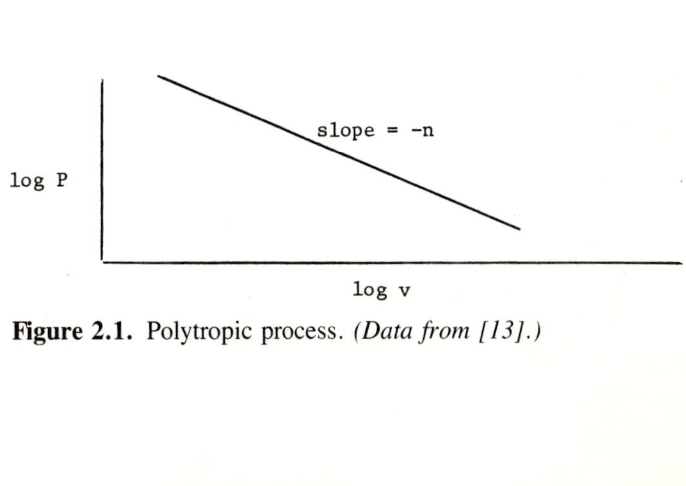

Figure 2.1. Polytropic process. *(Data from [13].)*

or

$$Pv^n = \text{constant} \tag{2.28}$$

where

$$n = \frac{\ln(P_2/P_1)}{\ln(v_1/v_2)} \tag{2.29}$$

With this, the following relations can be written:

$$\frac{P_2}{P_1} = \left(\frac{v_1}{v_2}\right)^n \tag{2.30}$$

$$\frac{T_2}{T_1} = \left(\frac{P_2}{P_1}\right)^{(n-1)/n} = \left(\frac{v_1}{v_2}\right)^{n-1} \tag{2.31}$$

ADIABATIC VERSUS POLYTROPIC PROCESS

An adiabatic process is a reversible constant entropy process for an ideal gas without heat transfer, following the relationship

$$Pv^k = \text{constant}$$

A polytropic process is a reversible process for an ideal gas with heat transfer, and variable entropy, following the relationship

$$Pv^n = \text{constant}$$

where

$n = 1$ for isothermal process (constant temperature)

$n = \infty$ for isometric process (constant volume)

All compressor processes fall between isentropic ($n = k$) and isometric ($n = \infty$).

Industry accepted practice is to use adiabatic equations for single-stage and air compressors, while polytropic equations are generally used for all other situations.

Remember:

- $H_{ad} < H_p$
- $\eta_{ad} < \eta_p$
- $GHP_{ad} = GHP_p$
- Summation of individual-stage adiabatic head (H_{ad}) does not equal the overall compressor adiabatic head.
- The summation of individual-stage polytropic head equals overall compressor polytropic head.

HEAD

For a reversible adiabatic process [14],

$$Pv^k = \text{constant}$$

$$P_1 v_1^k = Pv^k = \text{constant}$$

$$v = \left(\frac{P_1}{P_2}\right)^{1/k} v_1$$

$$H_{ad} = \int_{P_1}^{P_2} v\, dP$$

$$= \int_{P_1}^{P_2} (P_1/P)^{1/k} v_1\, dP$$

$$= \frac{k}{k-1} P_1 v_1 \left[\left(\frac{P_2}{P_1}\right)^{(k-1)/k} - 1\right]$$

Using $Pv = ZRT$,

$$H_{ad} = ZRT_1 \frac{k}{k-1} \left[\left(\frac{P_2}{P_1}\right)^{(k-1)/k} - 1\right] \tag{2.32}$$

For a polytropic process,

$$Pv^n = \text{constant}$$

$$\therefore\ H_p = ZRT_1 \frac{n}{n-1} \left[\left(\frac{P_2}{P_1}\right)^{(n-1)/n} - 1\right] \tag{2.33}$$

n = polytropic exponent

$$n = \frac{\log r_p}{\log r_v} = \frac{\ln r_p}{\ln r_v}$$

r_v = volume ratio = $v_1/v_2,\ r_v > 1$

$r_p = P_2/P_1$

$$\frac{n}{n-1} > 1$$

MOLLIER METHOD

When a Mollier chart of gas properties is available, the Mollier Method is used [11]. Here is how the head equation for this method is derived.

$$H_p = RT \frac{n}{n-1} \left[\left(\frac{P_2}{P_1}\right)^{(n-1)/n} - 1\right] = R\left(\frac{n-1}{n}\right)(T_2 - T_1)$$

$$= (P_2 v_2 - P_1 v_1)^{(n-1)/n}$$

but

$$\left(\frac{P_2}{P_1}\right)^{(n-1)/n} = \frac{T_2}{T_1} = \frac{P_2 v_2}{P_1 v_1}$$

or

$$\frac{n-1}{n} = \frac{\ln (P_2 v_2/P_1 v_1)}{\ln(P_2/P_1)}$$

or

$$\frac{n}{n-1} = \frac{\ln(P_2/P_1)}{\ln(P_2 v_2/P_1 v_1)}$$

Therefore

$$H_p = \ln\left(\frac{P_2}{P_1}\right)\left[\frac{P_2 v_2 - P_1 v_1}{\ln(P_2 v_2/P_1 v_1)}\right] \tag{2.34}$$

The last half of the equation is a log mean. Substituting an arithmetic mean introduces an insignificant error.

Therefore

$$H_p = \left[\ln \frac{P_2}{P_1}\right]\left(\frac{P_1 v_1 + P_2 v_2}{2}\right) \tag{2.35}$$

This can be rearranged to

$$H_p = 72\left[\ln \frac{P_2}{P_1}\right](P_1 v_1 + P_2 v_2) \tag{2.36}$$

$$= 166\left[\log \frac{P_2}{P_1}\right](P_1 v_1 + P_2 v_2) \tag{2.37}$$

Adiabatic head can be determined from a Mollier diagram by using the relationship

$$H_{ad} = h_{2_{ad}} - h_1 \tag{2.38}$$

Note: $h_{2_{ad}}$ is found by following a constant entropy line from point 1 to point 2. (See Figure 2.2.)

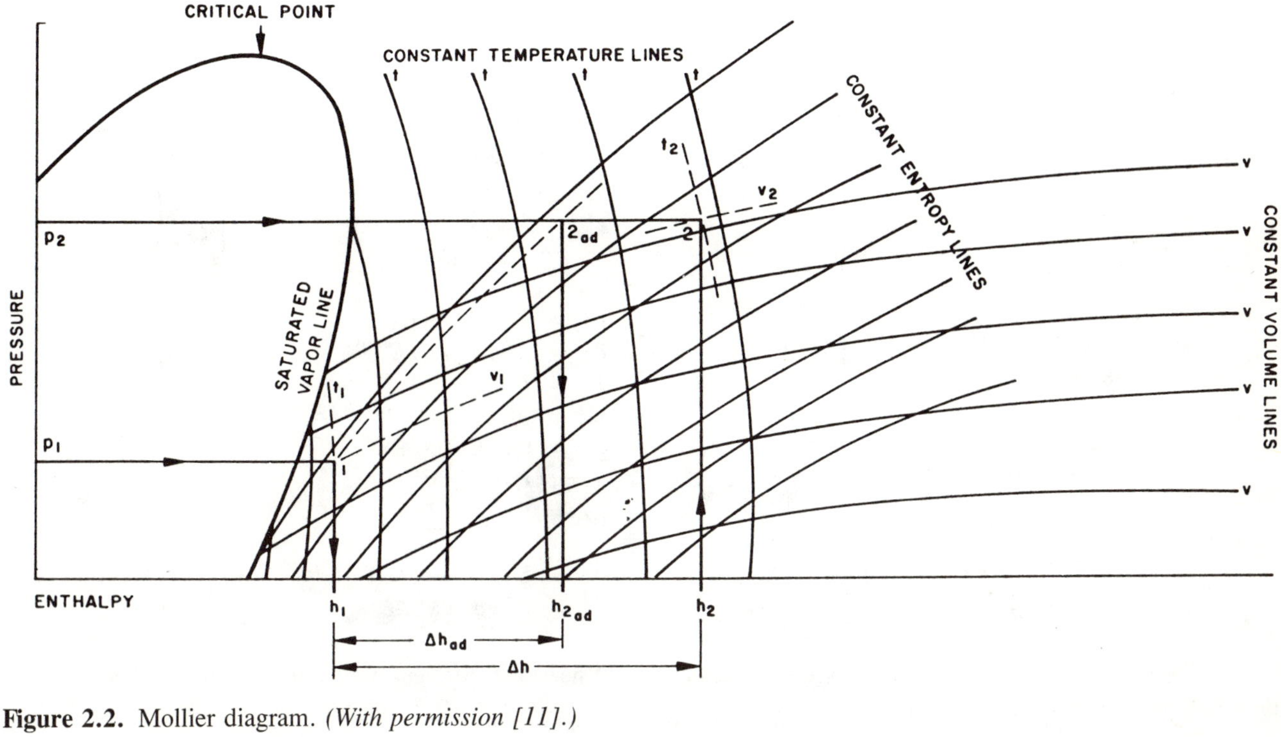

Figure 2.2. Mollier diagram. *(With permission [11].)*

CONCEPTUALIZING HEAD

Head = Energy
= Enthalpy
= Foot-pound force per pound mass, or $ft\text{-}lb_f/lb_m$

For a constant speed only CFM and head properly describe a particular compressor. Head is a function of speed and impeller geometry (diameter, tip opening, blade angle, etc.) and is not related to type of gas, temperature, or pressure. Discharge pressure for a given head is a function of gas mole weight, pressure, and temperature.

Imagine two incompressible gas columns, both 10,000 ft. high, both at atmospheric pressure and temperature at the uppermost ends, one column filled with air (MW = 29) and the other with butane vapor (MW = 58). For both cases, compressor geometry and speed are identical (Figure 2.3). Since the gas properties are different though, the discharge pressures for the two machines are significantly different, while the head (10,000 ft.) is the same.

This demonstrates that for a given head, a higher pressure ratio is obtained for a higher density gas.

WORK AND EFFICIENCY

The efficiency of a thermodynamic system is the ratio of work output of the system (head) to the work input to the system (shaft power).

The difference between head and work is the amount of losses internal to the machine due to such things as friction and windage. These losses show up as heat and add to the discharge temperature. These losses include

External to the Main Flow Path

Unavailable energy added

- windage and disk friction
- leakage

Internal to the Main Flow Path

Actual losses of blade input energy:

- skin friction
- blade loading and diffusion
- incidence angle
- exit mixing losses
- clearance losses

WORK

Using the General Energy Equation (first law of thermodynamics),

$$q + u_1 + P_1 v_1 + \frac{V_1^2}{2g} + Z_1 = W + u_2 + P_2 v_2 + \frac{V_2^2}{2g} + Z_2 \qquad (2.15)$$

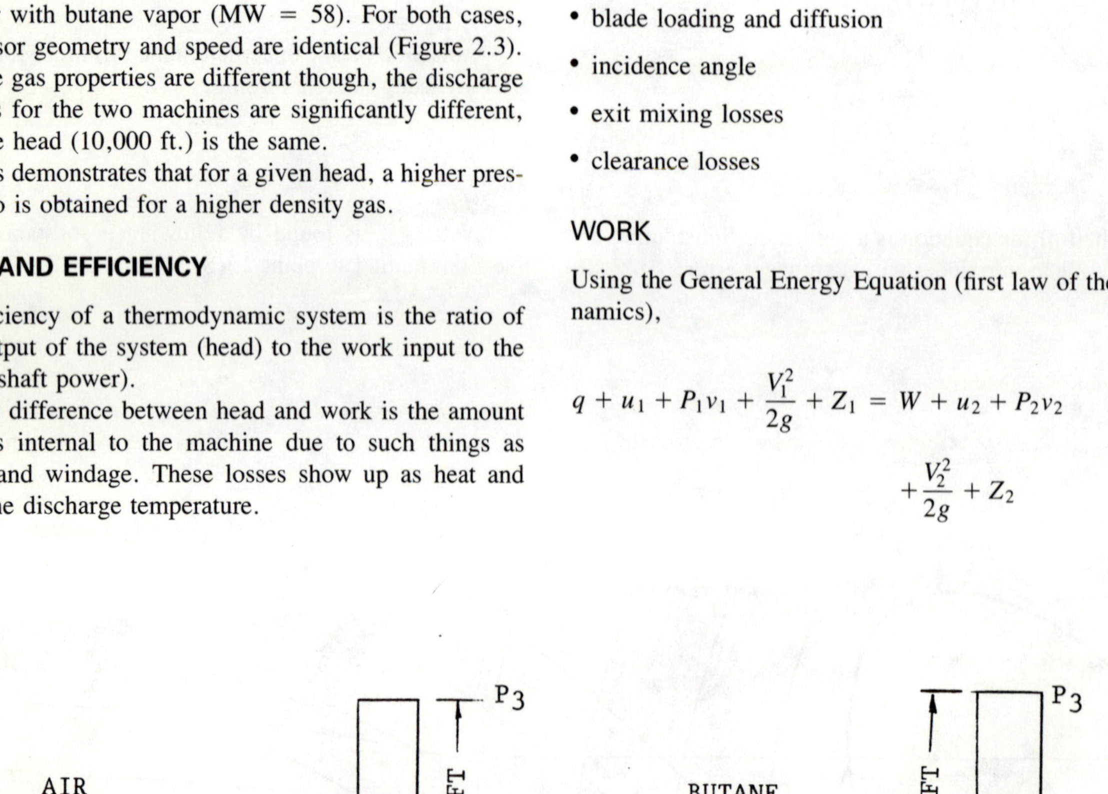

Figure 2.3. Two identical compressors provide the same head but different discharge pressures. *(Data from [10,14].)*

This can be reduced to

$$q + h_1 + \frac{V_1^2}{2g} = W + h_2 + \frac{V_2^2}{2g}$$

Assuming negligible heat transfer and minimal velocity effects,

$$W(\text{work}) = (h_2 - h_1) \tag{2.16}$$

For an isentropic process,

$$W = c_p(T_2 - T_1) \tag{2.39}$$

ADIABATIC EFFICIENCY

Adiabatic efficiency uses isentropic relationships to define head (useful work) and total work input.

$$\eta_{ad} = \frac{H_{ad}}{W} = \frac{\Delta h_{\text{isentropic}}}{\Delta h_{\text{actual}}} = \frac{c_p(T_{2_{ad}} - T_1)}{c_p(T_2 - T_1)} \tag{2.40}$$

where $T_{2_{ad}}$ represents isentropic compression (see Figures 2.4 and 2.2).

$$\eta_{ad} = \frac{T_{2_{ad}} - T_1}{T_2 - T_1}$$

$$\eta_{ad} = \frac{\Delta T_{\text{isentropic}}}{\Delta T_{\text{actual}}} \tag{2.41}$$

$$\frac{T_2}{T_1} = \left(\frac{P_2}{P_1}\right)^{(k-1)/k}$$

$$T_2 - T_1 = T_1\left(\frac{T_2}{T_1} - 1\right) = T_1[(r_p)^{(k-1)/k} - 1]$$

This gives

$$\eta_{ad} = \frac{T_1[(P_2/P_1)^{(k-1)/k} - 1]}{T_2 - T_1} \tag{2.42}$$

Another form of the Adiabatic Efficiency Equation [9]:

$$\eta_{ad} = \frac{RT_1 \ln(P_2/P_1)}{c_p(T_2 - T_1)} \tag{2.43}$$

The overall adiabatic efficiency is useful as a measure of the overall performance of a turbocompressor for the purpose of determining power. However, it is not always a true indication of efficiency regarding internal losses. Figure 2.4(a) illustrates this point. Since isentropic work is proportional to temperature rise ($W_{ad} = c_p\Delta T$), the distance from point 1 to point 2_{ad} is proportional to the adiabatic work required to compress the gas from P_1 to P_2. The actual work, however, is proportional to the vertical distance from point 1 to point 2.

For the three-stage compressor shown in Figure 2.4(c), note that this variation is further amplified. Also, note that the vertical distance (work) for the individual states W_a, W_b, plus W_c is greater than W_{ad} (total adiabatic work). This is due to the fanning effect of the pressure lines. $W_{ad} < (W_a + W_b + W_c) < W_p(1-2)$. Consider a compressor with an infinite number of stages. Then $W_{ad} < (W_1 + W_2 + W_3) + \cdots + W_\infty) = W_p(1-2)$. The following polytropic efficiency equation can be developed.

$$\eta_p = \frac{\ln(T_1/T_2)}{[(k-1)/k]\ \ln(P_1/P_2)} = \frac{n/(n-1)}{k/(k-1)} \tag{2.44}$$

This polytropic equation represents the true aerodynamic efficiency of a compressor for compression of an ideal gas [9]. There are, however, limitations to this equation also. Real gases do not always have a constant k value. The value of k for some gases at discharge conditions can vary significantly from the k value at suction conditions. The enthalpy (or Mollier) equation is the most accurate means of calculating the aerodynamic efficiency for any condition. In some cases, it is the only equation that can provide accurate results.

$$\eta_p = \frac{\text{Head}}{\text{Work input}} = \frac{H_p}{h_2 - h_1} \tag{2.45}$$

HORSEPOWER

Horsepower is the rate of doing work.

$$\text{HP} = \frac{\text{Work}}{\text{Time}} \tag{2.46}$$

$$\text{GHP} = W \times \dot{M} = (h_2 - h_1)\dot{M}$$

Since

$$\eta = \frac{H}{W}, \quad W = \frac{H}{\eta} \tag{2.47}$$

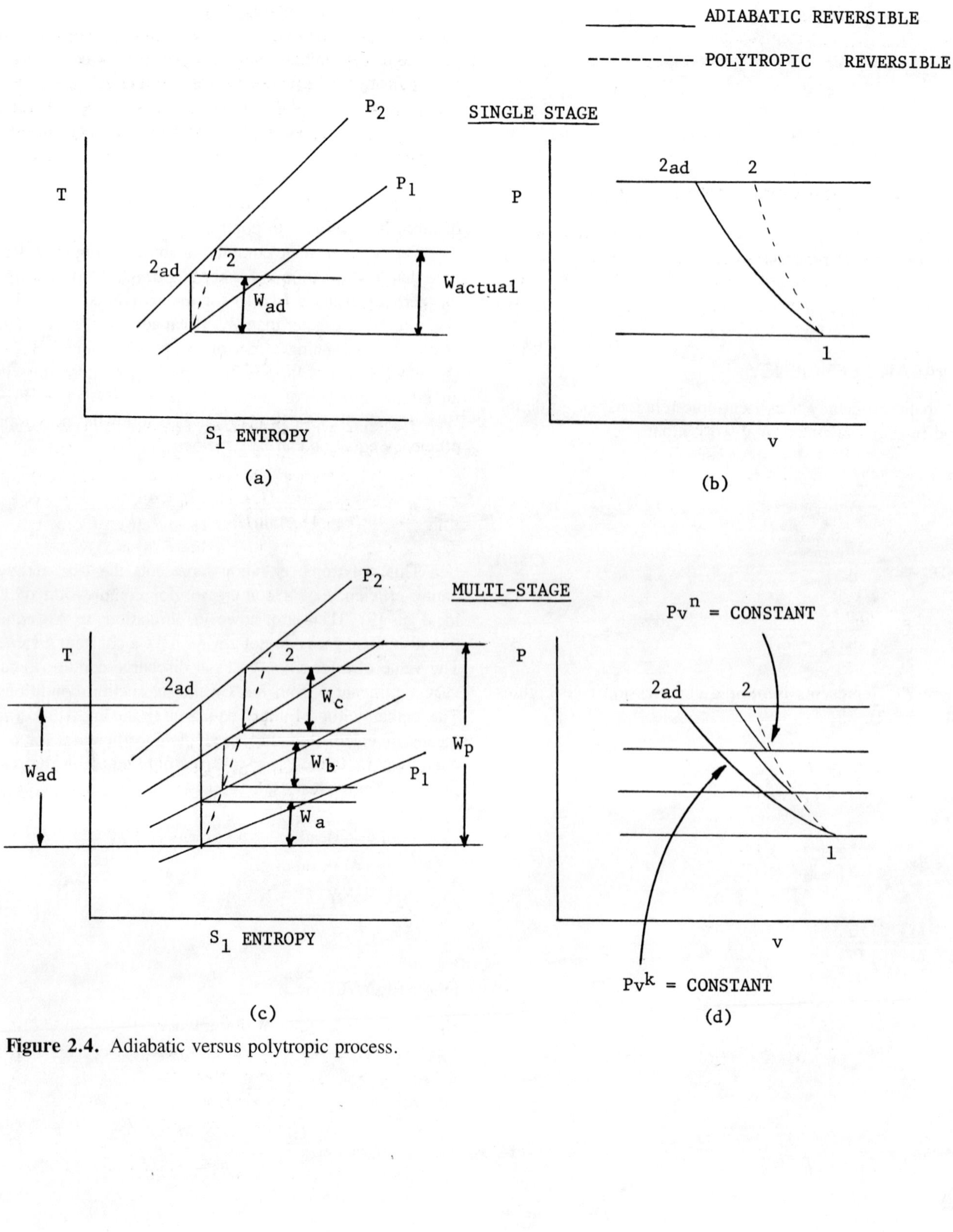

Figure 2.4. Adiabatic versus polytropic process.

Visualize a compressor lifting a weight of gas to a height (H) at a specified rate ($\dot{M}$).

$$\text{GHP} = \frac{H \times \dot{M}}{\eta} \tag{2.48}$$

FLOW MEASUREMENT, ORIFICE METERS

In order to derive the equation for flow through an orifice, we can use Bernoulli's equation with Station 1 being immediately upstream of the orifice and Station 2 at or immediately following the orifice [15].

$$P_1 v + \frac{V_1^2}{2g} = P_2 v + \frac{V_2^2}{2g}$$

$$V_2^2 - V_1^2 = 2gv(P_1 - P_2)$$

For an incompressible fluid,

$$Q_{s_1} = Q_{s_2} = VA, \quad V^2 = Q_s^2/A^2$$

Substituting:

$$\frac{Q_s^2}{A_2^2} - \frac{Q_s^2}{A_1^2} = 2gv\Delta P$$

$$Q_s = \frac{A_1 A_2}{\sqrt{A_1^2 - A_2^2}} \sqrt{2gv\Delta P}$$

$$= \frac{A_2}{\sqrt{1 - (A_2/A_1)^2}} \sqrt{2gv\Delta P}$$

$$A = \frac{\pi d^2}{4} \Rightarrow \frac{A_2}{A_1} = \left(\frac{D}{d}\right)^2$$

$$\beta = \frac{D}{d} = \text{Beta Factor}$$

$$Q_s = A_2 \frac{1}{\sqrt{1 - \beta^4}} \sqrt{2gv\Delta P}$$

$$Q_s = \frac{\pi d}{4} \frac{1}{\sqrt{1 - \beta^4}} \sqrt{2gv\Delta P}$$

where
d = orifice diameter, inches
D = pipe diameter, inches

$$\frac{1}{\sqrt{1 - \beta^4}} = \text{Velocity of Approach Factor, } E \tag{2.49}$$

Ideally, d should be the diameter of the vena contracta, but this value is unknown, or rather cannot be directly measured. Instead, a correction or efficiency factor is used. This factor is best determined by calibration, but can be calculated (see Chapter 9). This factor is referred to as the Discharge Coefficient, C. Incorporating this into the flow equation,

$$Q_s = \frac{\pi dCE}{4} \sqrt{2gv\Delta P} \tag{2.50}$$

C and E are sometimes combined and called the Discharge Coefficient, K.

$$Q_s = \frac{\pi dK}{4} \sqrt{2gv\Delta P} \tag{2.51}$$

The ratio of the differential pressure, ΔP, to the upstream static pressure, P, will influence the value of the overall flow coefficient. This occurs because there is a slight density change as the gas passes through the orifice meter. This effect is compensated for by the expansion factor, Y. The factor can be calculated or taken from a table or chart (see Figure 9.4).

We now have the flow equation

$$Q_s = \frac{\pi dKY}{4} \sqrt{2gv\Delta P} \tag{2.52}$$

Finally, to compensate for the thermal expansion of the orifice plate, the thermal expansion factor Fa is used. Generally this factor can be assumed to be 1.00 at moderate temperatures unless extreme accuracy is necessary (see Figure 9.3).

Including the thermal expansion effects,

$$Q_s = \frac{\pi dKY\text{Fa}}{4} \sqrt{2gv\Delta P} \tag{2.53}$$

GAS MIXTURES

For gases that do not significantly vary from the ideal gas laws, the following procedures may be used for calculating k (specific heat ratio and adiabatic exponent) and the compressibility factor z (correction for deviation from ideal gas law, $Pv = ZRT$). For gases that significantly vary from the ideal gas laws, such as heavy hydrocarbons, it will be necessary to use other more complex equations of state such as Benedict, Webb & Rubin (BWR), Lee Kesler, or others which are not covered here [11]. See the software suggested at the back of this book and the additional reading section for sources.

The properties of a gas mixture required for adiabatic compressor calculations are

1. Gas constant (dependent on molecular mass *MW*)
2. k, specific heat ratio and adiabatic exponent
3. P_1, T_1, v_1, and P_2
4. Compressibility, Z
5. Critical pressure, P_c
6. Critical temperature, T_c

Of the above properties of a gas mixture, MW, c_p, c_v, P_c, and T_c are calculated by adding the products of the individual mol fractions of each constituent, times its specific property (see Table 2.1). The temperature of any constituent is obviously the temperature of the mixture. The v (specific volume) of the mixture is obtained from $Pv = ZRT$. The compressibility of a mixture is obtained from a compressibility chart using the calculated values of P_c and T_c of the mixture. (See Equation (2.45). Polytropic efficiency calculations for heavy hydrocarbon gases require determination of enthalpy.)

The k value of a mixture is determined from

$$Mc_p - Mc_v = 1.985 \text{ BTU/(lb-mole-°R)} = 8.314 \text{ kj/kmol-°K}$$

$$k = c_p/c_v$$

or

$$k = \frac{Mc_p}{Mc_v} \text{ where M = molecular weight of the gas}$$

$$= \frac{Mc_p}{Mc_p - 1.985} \text{ where } Mc_p = \text{MW} \times c_p, \text{ or molal } c_p$$

For a gas mixture, the summation of the mole fraction times the molal c_p of each constituent is used (Table 2.1).

$$k = \frac{\Sigma Mc_p(\text{mix})}{\Sigma Mc_p(\text{mix}) - 1.985} \tag{2.54a}$$

or for metric values

$$k = \frac{\Sigma Mc_p(\text{mix})}{\Sigma Mc_p(\text{mix}) - 8.32} \tag{2.54b}$$

The compressibility (Z_1) of the mixture can be determined by finding the reduced temperature T_{R1} and the reduced pressure P_{R1} (using Table A.1) as follows

$$T_{R1} = \frac{T_1}{T_{c(\text{mix})}} \tag{2.55}$$

$$P_{R1} = \frac{P_1}{P_{c(\text{mix})}} \tag{2.56}$$

Then these values are entered on Figure A.1 (in Appendix A) to find Z.

MOLLIER DIAGRAMS

Compressor performance cannot be accurately predicted without detailed knowledge of the behavior of the gas or gases involved.

Mollier diagrams, of course, are readily available for most pure gases at "conventional" pressures and temperatures. However, in cryogenic areas or at very high pressure, some gas behaviors are difficult to predict. Gas properties in these areas therefore have been estimates determined through rather empirical methods.

TABLE 2.1 Procedure for Calculating Values for a Gas Mixture following Perfect Gas Laws

Gas Mixture	(1) Mol% each gas	(2) Mols/hr each gas	(3) Mol Mass	(4) (1) × (3)	(5) Mass %	(6)* T_c	(7)* P_c	(8) (1) × (6)	(9) (1) × (7)	(10)* Mcp	(11) (1) × (10)
......	......	......	......	a	a/d × 100	......	......	......	......	......	......
......	......	......	......	b	b/d × 100	......	......	......	......	......	......
......	......	......	......	c	c/d × 100	......	......	......	......	......	......
				d				......	......		......
Calculate $k_{(\text{mixture})} = \frac{\Sigma Mcp_{(\text{mix})}}{\Sigma Mcp - 1.985}$				Apparent Mol Mass of Mixture				$P_{c(\text{mix})}$	$T_{c(\text{mix})}$		ΣMcp

*See Table A.1 for items 6, 7, and 10.

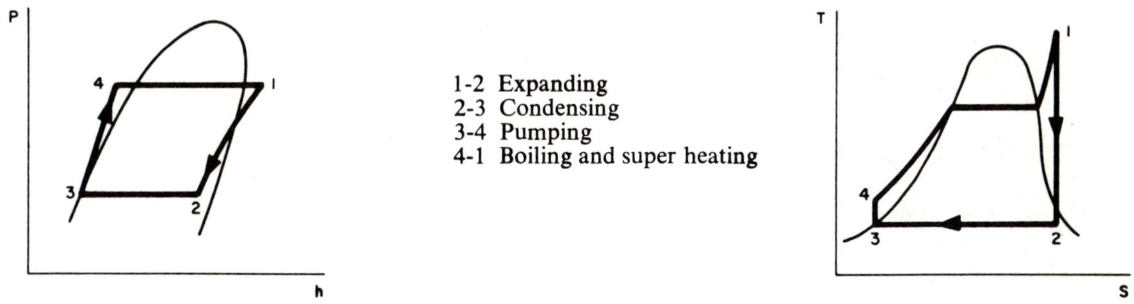

Figure 2.5. Rankine cycle. *(With permission [11].)*

The behavior of a wide variety of gases—in any conceivable mixture—can be accurately computed, plotted, and offered to the process engineer in the form of a Mollier diagram [11].

The only input required to obtain a plot of gas behavior is the identity and proportion of the gases involved (if a gas mix), and the limiting pressure and temperature values.

From the Mollier diagram, enthalpy and specific volume can be determined directly if the pressure and temperature of the gas are known. This enables very accurate calculation of head, efficiency, and flow.

Temperature Entropy (T-S) and Pressure Enthalpy (P-h) curves are commonly used to display gas properties. While these curves are primarily used for steam or refrigeration cycles, the curves are very useful for any process. The process is easy to "visualize" when plotted on a Mollier diagram. One can "see" the phase change, the expansion, or the compression process, and therefore easily comprehend the overall process and the effect of process changes.

STEAM GENERATION

A very common cycle depicted in Figure 2.5 is the Rankine cycle of the steam generating plant.

REFRIGERATION

The reverse of the Rankine or Carnot cycle is the vapor compression or refrigeration cycle (Figure 2.6).

Economizer cycles are commonly used to improve system efficiency (Figure 2.7).

Refrigeration is second only to air compression as the most common application of centrifugal compressors.

In the early years of the twentieth century most mechanical refrigeration capacity was used to produce manufactured ice for the preservation of food. These units used gases such as ammonia and carbon dioxide. Gradually the scope was increased until today refrigeration is applied in many areas other than the home, commercial, and comfort fields for which the flouridated hydrocarbons are used extensively.

Refrigeration is used in such processes as the manufacture of ethylene, ammonia, alkylation, de-waxing, etc. Most industrial refrigeration is an area associated with oil refining and therefore uses the readily available hydrocarbons as refrigerants.

The use of hydrocarbons as refrigerants has advantages other than availability. They have high mol weights, which permit fewer wheels to reach condensing pressure.

Figure 2.6. Carnot cycle. *(With permission [11].)*

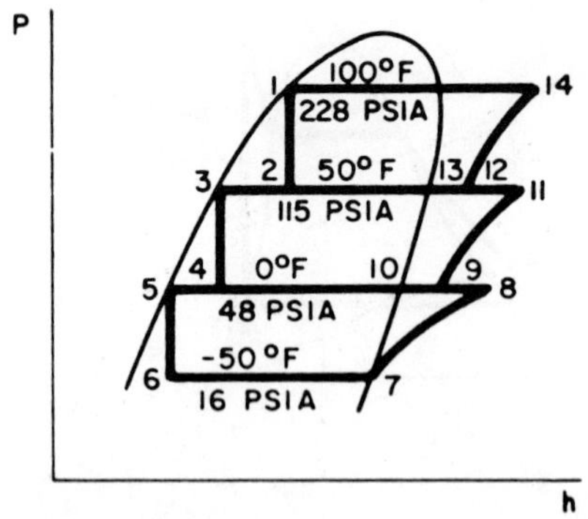

Figure 2.7. Economizer cycle. *(With permission [11].)*

Also, there is a sufficient variety of hydrocarbons to provide a relatively good choice of refrigerants in the temperature range from −300°F to +60°F.

The most commonly used hydrocarbon refrigerants are listed below:

Gas	Cooling Temperature Range
Methane	−300° to −200°F
Ethylene and Ethane	−175° to − 75°F
Propylene and Propane	− 50° to 0°F
Butane	+ 10° to + 60°F

Refrigeration capacity refers to the total amount of heat absorbed in the cooler. This is generally referred to in tons of refrigeration or more simply Tons.

The unit of Ton came about from the early days of cooling, in which ice was used in almost all cases. The natural amount of "cold" or "coldness" was a given weight of ice. A meaningful amount was a Ton. Therefore, the definition of a ton of refrigeration, or just a Ton, is the amount of heat absorbed by a ton of ice at 32°F melting to a ton of water at 32°F in a period of 24 hours. Since the latent heat of fusion of ice is 144 BTU/lb, a Ton, then, is equivalent to the absorption of 288,000 BTU/day or 12,000 BTU/hr or 200 BTU/min.

PERFORMANCE COEFFICIENTS

In order to condense the large number of variables in the study and design of aerodynamic components, dimensionless coefficients are used extensively. The coefficients are developed via Buckingham's Pie theorem, where repeating variables are systematically combined to form dimensionless groups. Some of the useful dimensionless groups plus some other frequently used values follow.

Reynolds Number

$$\text{Rey \#} = \frac{\text{Inertial Forces}}{\text{Viscous Forces}} = \frac{\text{Momentum}}{\text{Friction}} = \frac{VD}{\nu'} \tag{2.57}$$

$$\frac{VD\rho}{\mu'} \tag{2.58}$$

Mach Number

$$\text{Mach \#} = \frac{\text{Inertial Forces}}{\text{Compressibility Forces}} = \frac{\text{Momentum}}{\text{Elasticity}}$$

$$= \frac{V}{a} \tag{2.59}$$

Flow Coefficient

$$\frac{700}{Nd^3}Q = \phi \tag{2.60}$$

Head Coefficient

$$\frac{H}{U^2/g} = \mu \tag{2.61}$$

Work Coefficient

$$\frac{W}{U^2/g} = \gamma \tag{2.62}$$

Efficiency

$$\eta = \frac{\mu}{\gamma} \tag{2.63}$$

where

Q = Volume Flow CFM
N = Speed, RPM
d = Impeller Dia. In.
H = Head ft-lb_f/lb_m
U = Tip Speed, FPS

$$U = \frac{\pi dN}{720} = \frac{dN}{229.18}$$

g = Gravitational Constant 32.2 ft-lb_m/lb_f-sec^2
W = Work = $h_2 - h_1$ ft-lb_f/lb_m

Equivalent Tip Speed, FPS

$$U\sqrt{\theta}$$

where

$$\sqrt{\theta} = \sqrt{\frac{26.2MW}{kTZ}} = \frac{\text{Speed of Sound—Air}}{\text{Speed of Sound—Gas}} \quad (2.64)$$

Specific Speed

$$N_s = \frac{N\sqrt{60Q}}{H^{.75}} \quad (2.65)$$

FAN LAWS

From these relationships off-design performance can be approximated. These relationships are most accurate for single-stage compressors and an ideal gas. For multi-stage compressors with gas that deviates from the ideal gas laws, the accuracy of the following equations is reduced. Equations (2.66) and (2.67) are graphically represented in Figure 2.8.

$$Q \propto N \quad (2.66)$$

$$H \propto N^2 \quad (2.67)$$

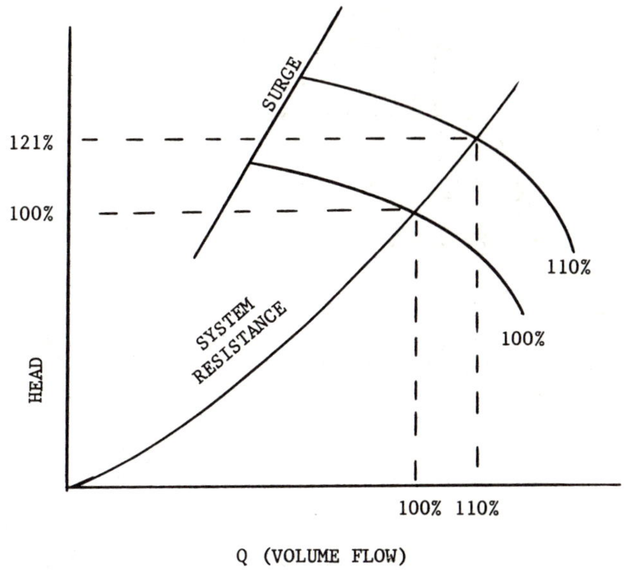

Figure 2.8. Fan laws. *(Data from [10,11].)*

$$\ln r_p \propto N^2 \quad (2.68)$$

$$(T_2 - T_1) \propto N^2 \quad (2.69)$$

$$\text{GHP} \propto N^3 \quad (2.70)$$

3

AERODYNAMIC COMPONENTS

The design of the primary components of both axial and centrifugal compressors is relatively simple. Any complexities arise in trying to get everything to work together with a high degree of efficiency—mechanically as well as aerodynamically. This chapter describes those main components, explains their functions, and covers some areas that need special consideration.

AXIAL COMPRESSORS

The major components of an axial compressor consist of (a) inlet nozzle, (b) prewhirl vanes, (c) rotating vanes, (d) stator vanes, (e) dewhirl vanes, and (f) discharge nozzle (Figures 3.1 and 3.2).

The nozzle guides and accelerates the gas stream into the prewhirl vanes which turn the gas stream to properly align it with the rotating blades. While both the rotating and stationary vanes act as diffusers, the rotating vane's primary function is to add to the total energy of the gas stream. Stator vanes, acting both as diffusers and reversing blades, orient the flow properly for the next row of rotating blades. While the first few rows of stator vanes are generally adjustable to compensate for off-design conditions, most rows of stator vanes are fixed. The dewhirl vanes remove the swirl from the gas stream before it enters the diffuser section.

CENTRIFUGAL COMPRESSORS*

The major elements of the centrifugal compressor (Figure 3.3) consist of (a) the inlet nozzle, (b) inlet guide vanes, (c) impeller, (d) radial diffuser, (e) return channel, (f) collector volute, and (g) discharge nozzle.

The inlet nozzle accelerates the gas stream and directs it into the inlet guide vanes which may be fixed or adjustable.

On a multi-stage compressor, the inlet nozzle is generally radial. In this case the inlet guide vanes are necessary to properly distribute the flow evenly to the first-stage impeller (Figure 3.4). Single-stage compressors frequently incorporate an axial inlet. In this case, inlet vanes may not be necessary.

Simply stated, kinetic energy is imparted to the gas via the impeller by centrifugal forces. The diffuser then reduces the velocity and converts the kinetic energy to pressure energy (Figure 3.5).

Because of the rotational effects of the impeller, the gas travels through the diffuser in a spiral manner. Therefore, before entering the next impeller, the flow must be straightened out by the return channel vanes (Figure 3.6).

DIAPHRAGMS

A diaphragm consists of a stationary element which forms half of the diffuser wall of the former stage, part of the return bend, the return channel, and half of the diffuser wall of the later stage. Due to the pressure rise generated, the diaphragm (Figure 3.7) is a structural as well as an aerodynamic device.

For the last stage or for a single-stage compressor, the flow leaving the diffuser enters the discharge volute. It is common to design these volutes for constant angular momentum ($R_i V_i$ = constant). This generally results in some velocity change through the volute (V_2 to V_5). Once the gas leaves the volute, it passes through the discharge nozzle which reduces the velocity somewhat before entering the process piping. Figure 3.8 represents such a constant angular momentum volute.

Since velocities are relatively high through the diffuser section (several hundred feet per second), surface finish/friction factor is crucial to overall efficiency of the unit.

In many processes, dirt or polymer buildup on the diaphragm surfaces will give the aerodynamic surfaces a rough finish (Figure 3.9). In some cases, polymer buildup has been known to severely restrict the diffuser passage. Both conditions cause increased pressure losses and result in reduced overall efficiency of the compressor.

Surface finish in these critical areas can be enhanced and preserved by applying a non-stick coating such as fluorocarbon-based material, or a corrosion-resistant coating such as electroless nickel. The long-term effects are shown in Figure 3.10.

INTERSTAGE SEALS

Due to the pressure rise across successive compression stages, seals are required at the impeller eye and rotor shaft to prevent gas backflow from the discharge to inlet end of

*Adapted from "Compressor Components," M. Sassos, Elliott Company, Jeannette, PA, 1986, with permission [16].

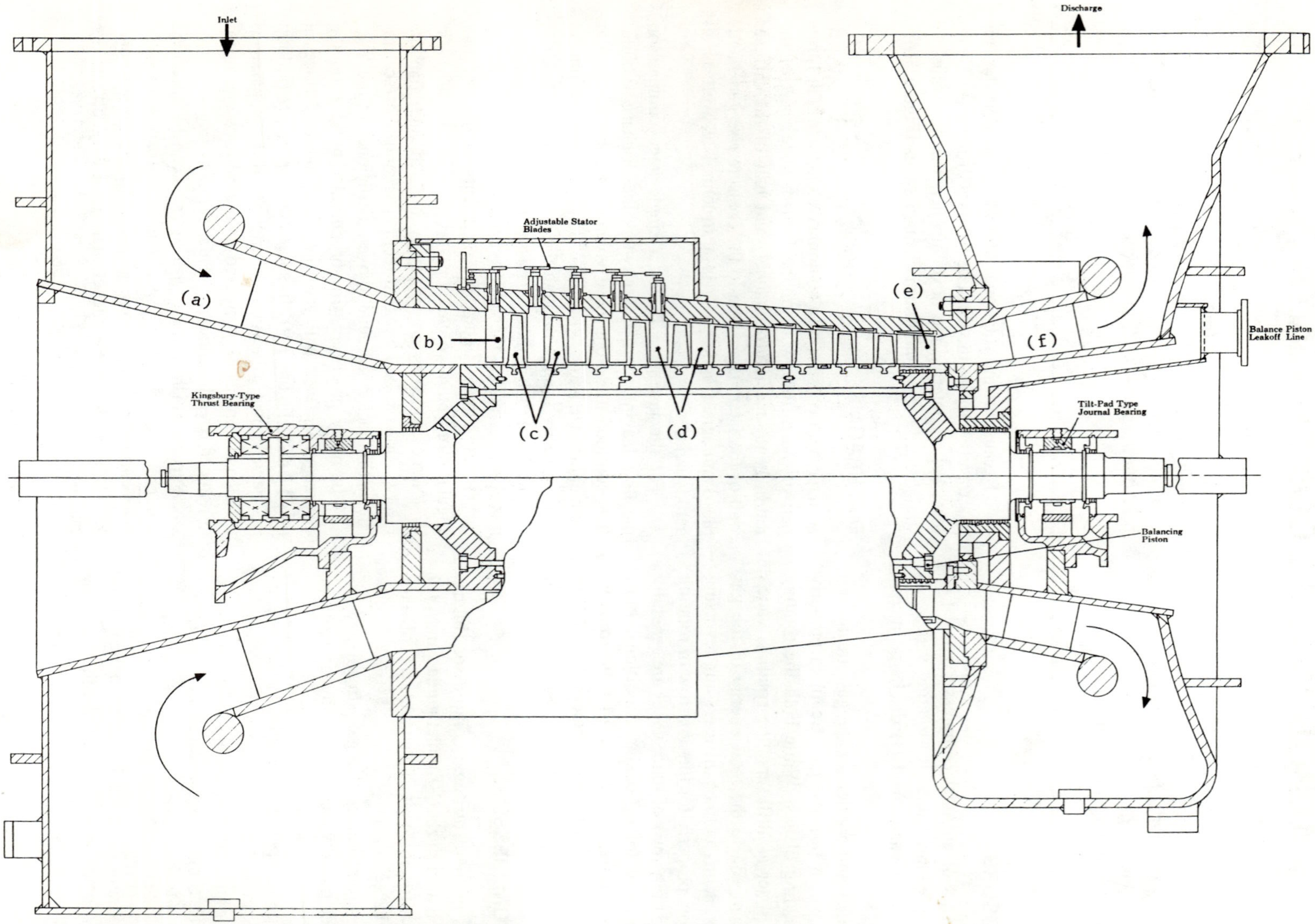

Figure 3.1. Axial compressor. *(Courtesy of Elliott Company, with permission.)*

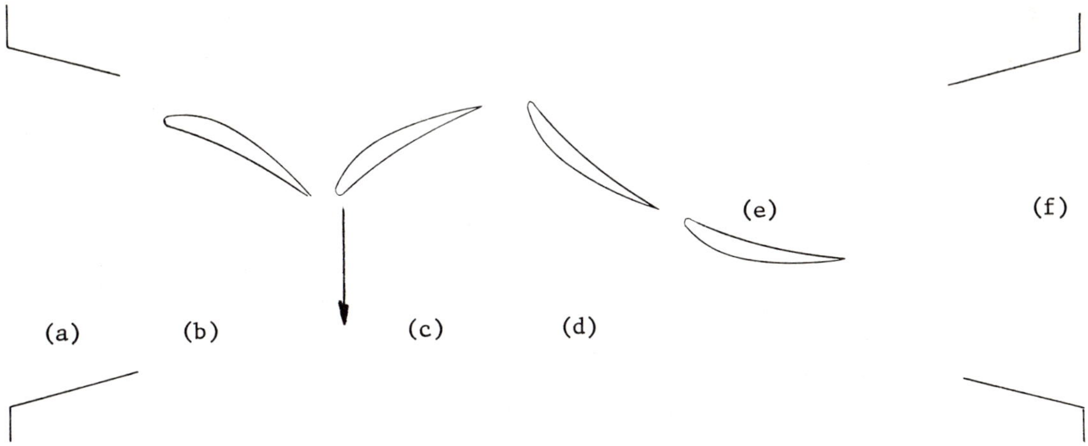

Figure 3.2. Axial components.

the casing. The condition of these seals therefore directly affects the compressor performance.

The simplest and most economical of all shaft seals is the straight labyrinth shown in Figure 3.11. This seal is commonly utilized between compression stages and consists of a series of thin strips or fins which are normally part of a stationary assembly mounted in the diaphragms. A close clearance is maintained between the rotor and the tip of the fins.

The labyrinth seal is equivalent to a series of orifices. Minimizing the size of the openings is the most effective way of reducing the gas flow. Labyrinths clogged with dirt

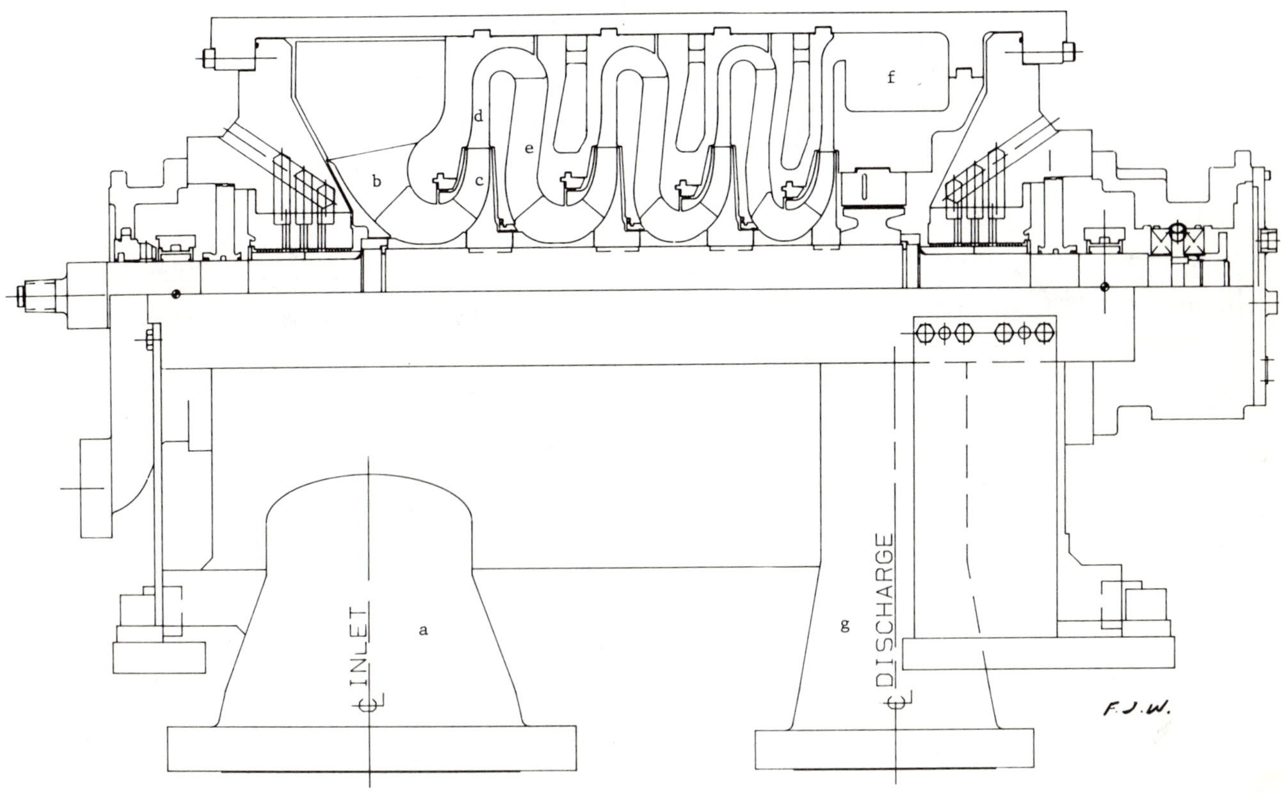

Figure 3.3. Multi-stage centrifugal compressor.

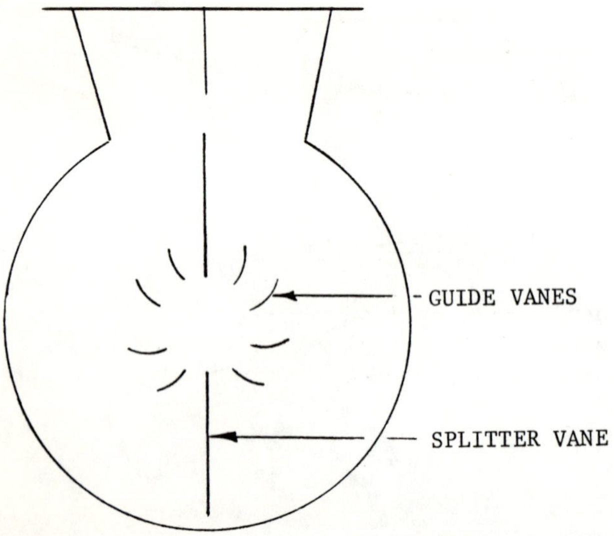

Figure 3.4. Multi-stage compressor inlet showing splitter vanes and guide vanes.

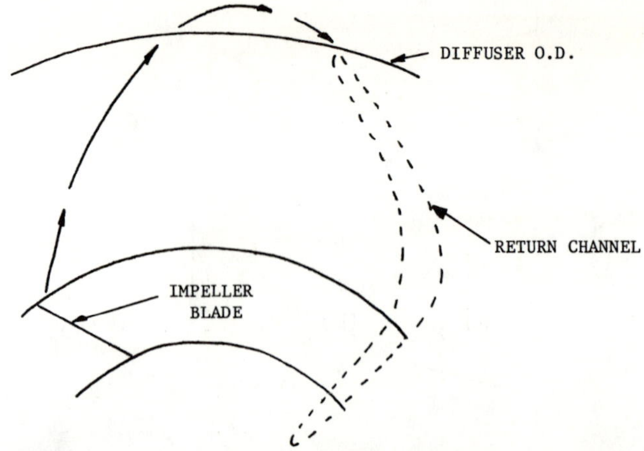

Figure 3.6. Flow path of gas from impeller tip to return channel.

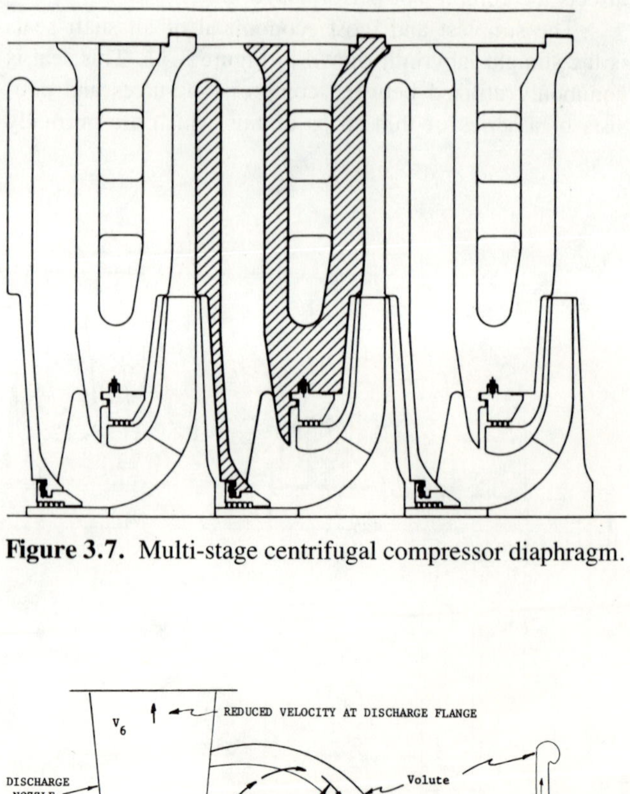

Figure 3.7. Multi-stage centrifugal compressor diaphragm.

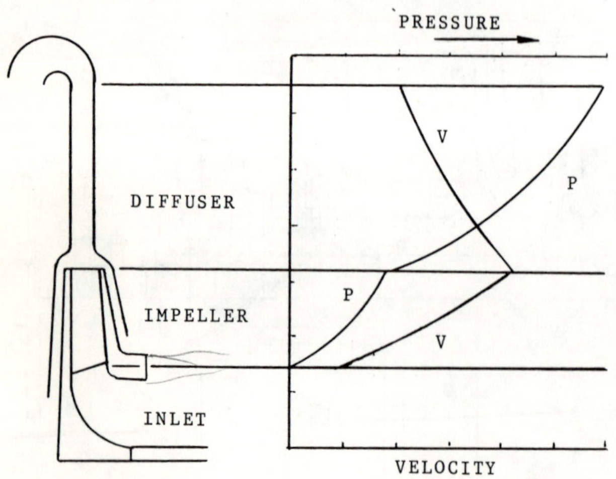

Figure 3.5. Velocity/pressure development. *(Data from [10].)*

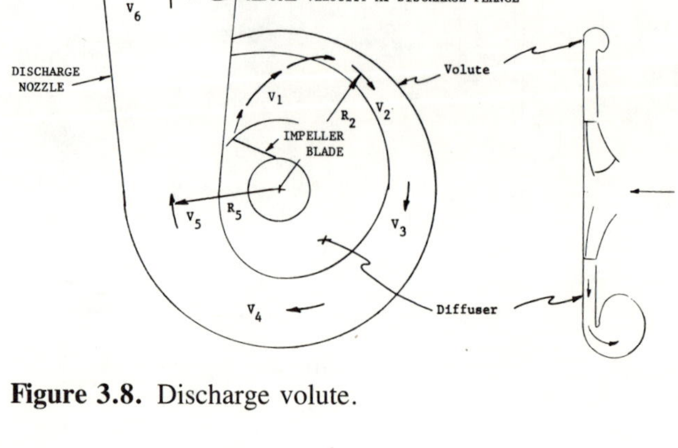

Figure 3.8. Discharge volute.

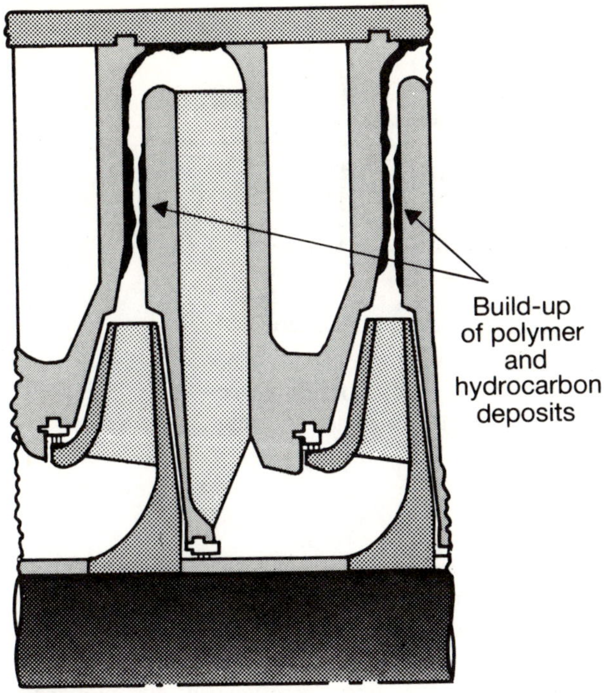

Figure 3.9. Dirt or polymer buildup in diffuser passages of a centrifugal compressor.

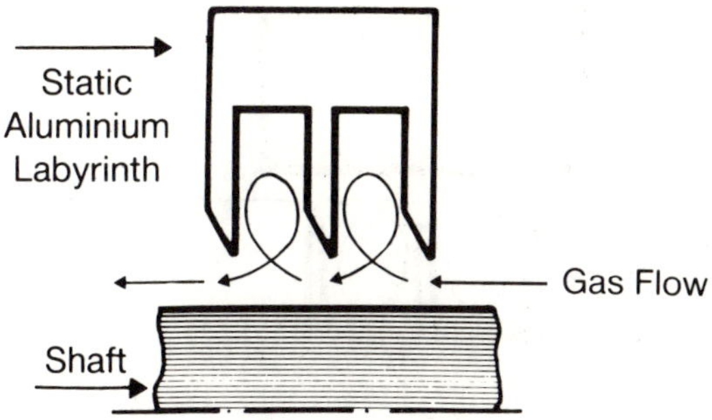

Figure 3.11. Aluminium labyrinth. New and clean. Turbulence creates resistance to leakage flow.

Figure 3.12. Fouled labyrinth. Turbulence reduced, leakage flow increased.

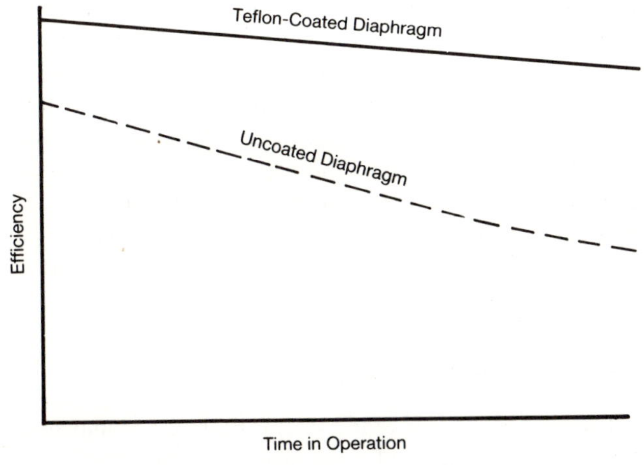

Figure 3.10. Effect of coated and non-coated surfaces on dirt/polymer buildup in diffuser passages of a centrifugal compressor.

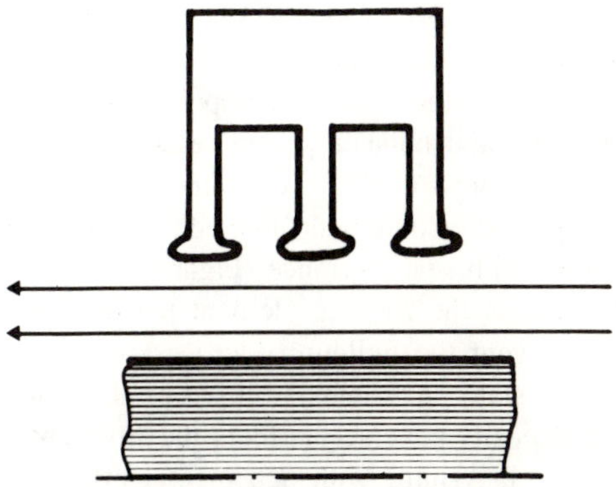

Figure 3.13. Rubbed labyrinth. Clearance increased, turbulence reduced, leakage increased.

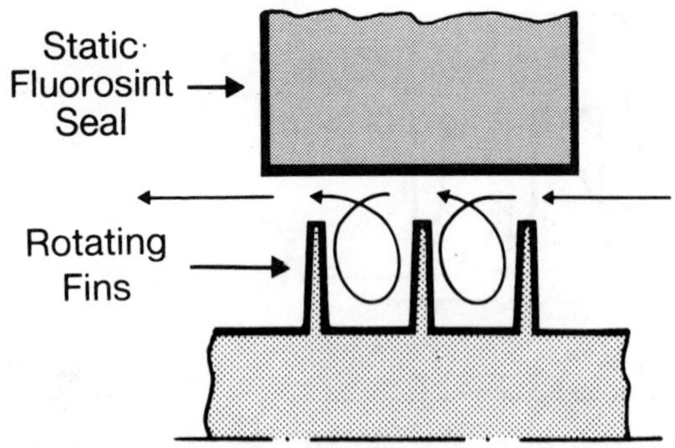

Figure 3.14. Abradable seal. Turbulence creates resistance to leakage flow.

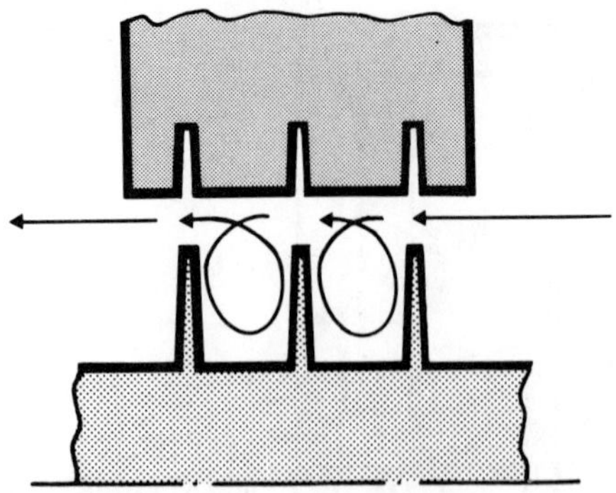

Figure 3.15. Rubbed abradable seal. Running clearance unaltered, turbulence continues to create resistance to leakage flow.

(Figure 3.12) and worn or wiped labyrinths with increased clearances (Figure 3.13) allow larger gas leakage. This can affect compressor operation, and therefore the seals should be replaced.

Labyrinth material has typically been aluminum, because aluminum is compatible with most gases and is ductile enough to prevent rotor damage in the event of rubbing. A hard labyrinth material could result in dry whirl and catastrophic failure of the compressor.

Calculations and field performance data indicate that wiped interstage seals can decrease unit efficiency by 7% or more. Operating modes that contribute to labyrinth damage include extended surging, prolonged running in the critical speed regions, and liquid slugging.

In order to reduce or negate the performance effects common with damaged interstage seals, several improvements have been adopted by compressor manufacturers in the past several years. Most noteworthy is the use of abradable seals in the impeller eye and shaft seal areas. Advantages include tighter design operating clearances and minimal efficiency effects after a seal rub, as shown in Figures 3.14 and 3.15.

The efficiency gain of abradable seals is achieved through the reduction of seal clearances, thereby reducing recirculating flow through the impellers. Impeller eye seals, interstage shaft seals, and balance piston seals are effective in improving compressor efficiency when changed to the abradable design. Abradable seals also control impeller thrust, which varies with seal clearance (Figure 3.17).

Having the fins as the rotating element permits centrifugal force to prevent the buildup of process deposits. Where conventional static labyrinths are used on a fouling duty, buildup of deposits adversely affects the flow characteristic across the labyrinth, with detrimental effect on compressor efficiency. Rotating fins minimize this problem. See Figures 3.11, 3.12, and 3.14.

A rub on an aluminum labyrinth causes the tips of the aluminum fins to mushroom out (Figure 3.13). This creates undesirable flow characteristics across the labyrinth and increases the radial clearance. These two factors are both detrimental to compressor efficiency and will have an effect on the thrust loading of the machine. With the abradable design, the rotating fins rub into the static element, without damage to the fins and without effect on the normal running clearances. No performance deterioration or change in thrust load occurs (Figures 3.14 and 3.15).

The overall efficiency improvement attainable by using abradable seals in a compressor varies with several factors, most notably the size of the compressor. Flow capacity increases as the square of the impeller diameter, while seal clearance increases more linearly with impeller size and is also dependent on other factors such as bearing clearances and manufacturing tolerances. Therefore, as the compressor size increases, the leakages involved become a smaller portion of the total flow. As this happens, the improvements gained by reducing these leakages have a diminishing impact on the machine's overall efficiency. Therefore, it is the smaller, higher pressure compressors that benefit most from abradable seal conversions.

BALANCE PISTON SEAL

A balance piston (or a center seal) is utilized to compensate for aerodynamic thrust forces imposed on the rotor due to the pressure rise through a compressor. The purpose of the balance piston is to utilize the readily available pressure differentials to oppose and balance most of these thrust forces. This enables the selection of a smaller thrust bearing, which results in lower horsepower losses.

A certain amount of leakage occurs across the balance piston since a labyrinth seal is utilized. This leakage, which is a parasitic flow (a horsepower loss) is normally routed

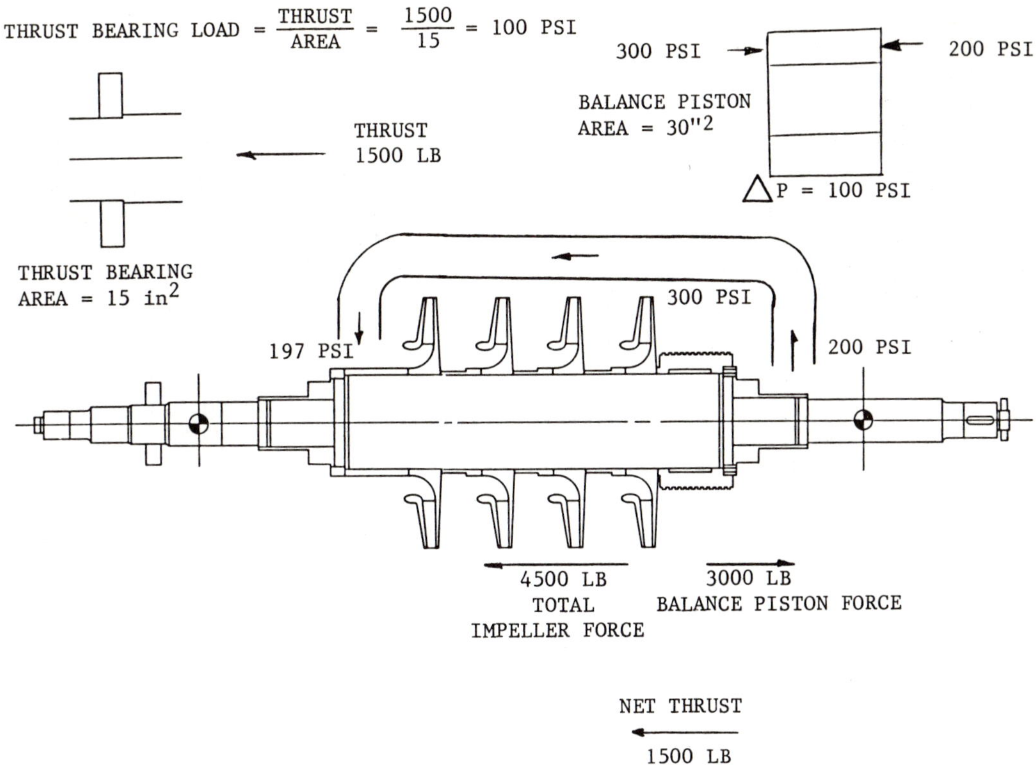

Figure 3.16. Schematic of compressor thrust. Pressure drop in balance line is normally 1 to 3 PSI. *(Data from [11].)*

back to the compressor suction, thus creating a known differential pressure across the balance piston (Figure 3.16). Occasionally, leakoff may be routed to other sections to gain an efficiency advantage. Air compressors generally route the balance piston leakage to atmosphere.

Since the balance piston seal must seal the full compressor pressure rise, integrity of this seal is crucial to good performance. A damaged seal results in higher leakage rates, higher horsepower consumptions, and greater thrust loads.

One user of a compressor noted the following data before and after a balance piston seal replacement. This machine was in refrigeration service and was required to maintain a constant discharge pressure.

		Before	After
Discharge Pressure	(PSIG)	410	410
Discharge Temperature	(°F)	142	116
Axial Position	(Mils)	24	19
Balance Line ΔP	(PSID)	4.7	1.5
Speed	(RPM)	11440	10770
Thrust Metal Temperature	(°F)	240+	165

The balance piston damage was a result of surging and vibration excursions. The interstage seals were also extensively damaged, which contributed to the poor compressor efficiency. Note the differences in the various data for before and after the seal replacement. The discharge temperature was high, since more work input was required to achieve the desired discharge pressure. In order to get the higher level of work input, the speed was increased. The wiped seals not only caused increased inefficiencies, but also higher thrust loads. This showed up in the axial position and thrust bearing temperature.

IMPELLER THRUST*

Impeller thrust is generated by the differential force on the cover and hub of the wheel. These forces are the summation of the product of the pressures acting on the cover, hub, and the differential area from the shaft to the tip of the wheel.

---

*Adapted from "Compressor Refresher," Elliott Company, Jeannette, PA, 1975, with permission [11].

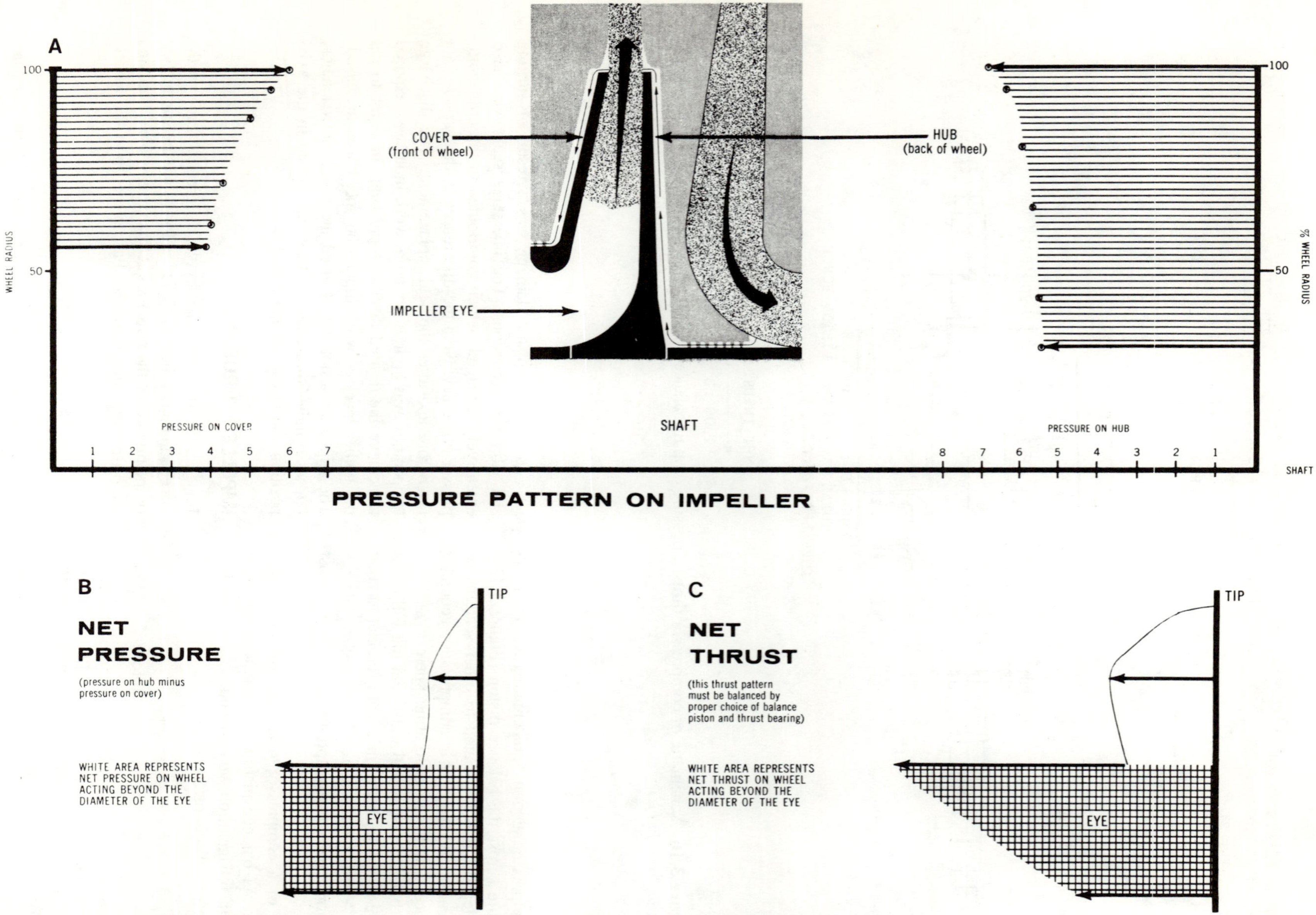

Figure 3.17. (A) Pressure pattern on impeller. (B) Net pressure. (C) Net thrust. *(With permission [11].)*

The impeller generates thrust between the eye and tip of the wheel, as well as below the eye. These forces (thrust) are caused by several different effects:

1. rotational inertia field
2. leakage
3. friction
4. diffusion

The effects of friction and diffusion are insignificant.

As indicated by the flow paths (Figure 3.17), gas will flow toward the tip of the wheel along the hub, and toward the eye of the wheel along the cover. Due to the pressure rise in the diffuser, the return channel pressure is greater than the pressure behind the impeller hub. Leakage therefore occurs from the return channel toward the impeller hub and outward toward the impeller tip. The effect of the pressure established by this leakage, superimposed upon the rotating inertia field, is shown in Figure 3.17A.

From Figure 3.17 it can be seen that there is an obvious net pressure differential toward the suction of the machine in addition to the area caused by the eye of the impeller. This is indicated in Figure 3.17B. Integrating the products of the pressure and area from eye to tip results in the net thrust on a wheel.

For a "perfect" seal at the impeller eye and shaft areas, the thrust is only a function of the area inside the eye seal. For a "real" seal with clearance and leakage in these areas, the net thrust is approximately 50% greater. As the clearances increase beyond the design values, thrust values increase even further.

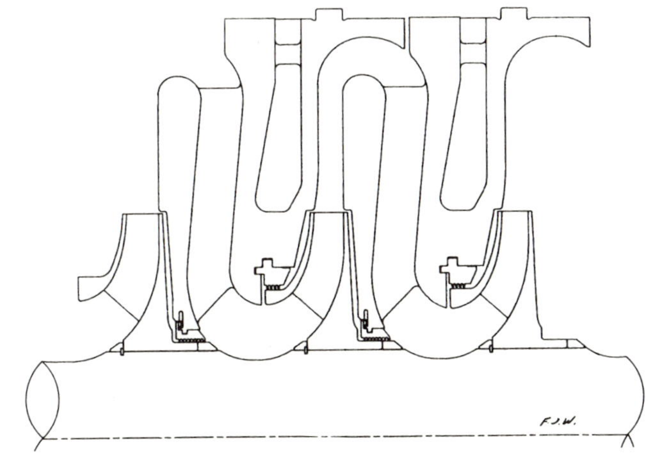

Figure 3.18. High efficiency centrifugal compressor stage.

EFFICIENCY IMPROVEMENTS

Increased costs for energy have resulted in improvements to the basic staging design. Features that were not economical 15 or 20 years ago can now be utilized as cost-effective means of reducing energy consumption.

Figure 3.18 is a good example of modern high-efficiency staging design. Every feature has been designed with high efficiency in mind. Computerized design enables precise flow path area control to minimize gas velocity changes. Full, gentle, radiused flow paths and three-dimensional blading minimize flow separation. Improved interstage shaft and impeller eye seals prevent recirculation of gas.

By including high-efficiency staging like Figure 3.18, abradable interstage and balance piston seals, and coated diffusers in a new or rebuilt compressor, polytropic efficiencies of 85% or more are feasible.

4

COMPRESSOR CHARACTERISTICS

As noted in Chapter 1, compressor performance is a function of the type of compressor. As well as the hardware configuration, the type of gas plays its part in determining the compressor characteristic curve. The effects of various parameters such as diffusor width, blade angle, gas density, compressor speed, and Mach Number in the development of the compressor characteristic curve shape are discussed in this chapter.

CENTRIFUGAL COMPRESSORS*

The characteristics of a centrifugal compressor (Figure 4.1) are determined by the impeller, diffuser, and return channel or volute geometry.

There are three important aspects of the compressor curve that will be discussed (Figure 4.2):

1. Slope of the curve
2. Stonewall (or choke)
3. Surge

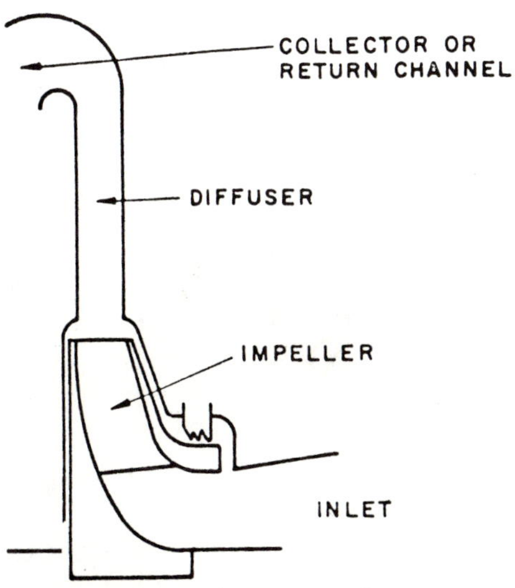

Figure 4.1. Centrifugal compressor stage [18].

*Adapted from "Centrifugal Compressors . . . The Cause of the Curve," D.C. Hallock, Elliott Company, Jeannette, PA, 1968, with permission [17]; and "Compressor Performance," R. Salisbury, Elliott Company, Jeannette, PA, with permission [18].

SLOPE

To understand about the slope of the centrifugal compressor head curve it is necessary to first understand what is going on at the impeller discharge in terms of velocity vector diagrams [18].

V_{rel} (Figure 4.3) represents the gas velocity relative to the blade. U_2 represents the absolute tip speed of the blade. The resultant of these two velocity vectors is represented by V, which is the absolute velocity of the gas ($U_2 + V_{rel} = V$). Knowing the magnitude and direction of this absolute velocity, we can break this vector into its radial and tangential components (Figure 4.4).

For a radial inlet impeller the head output is proportional to the product of U_2 and V_T.

For a typical backward-leaning bladed impeller, as the flow decreases at constant speed, V_{rel} decreases. This makes V_T increase, which increases head output. This head increase with decreasing flow is what causes the basic slope

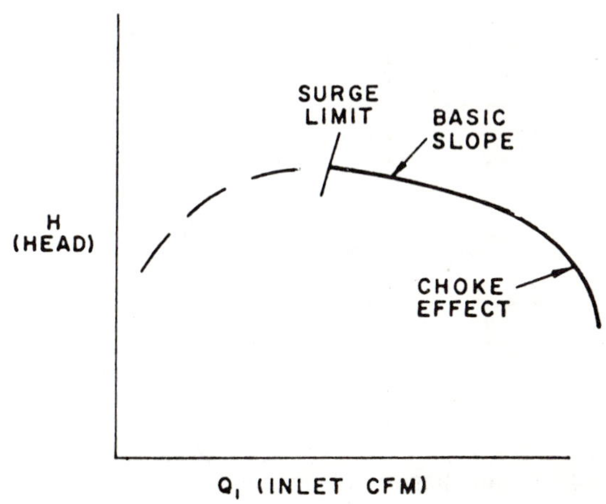

Figure 4.2. Head curve for a compressor stage [17]. *(Used courtesy of Elliott Company, Jeannette, PA.)*

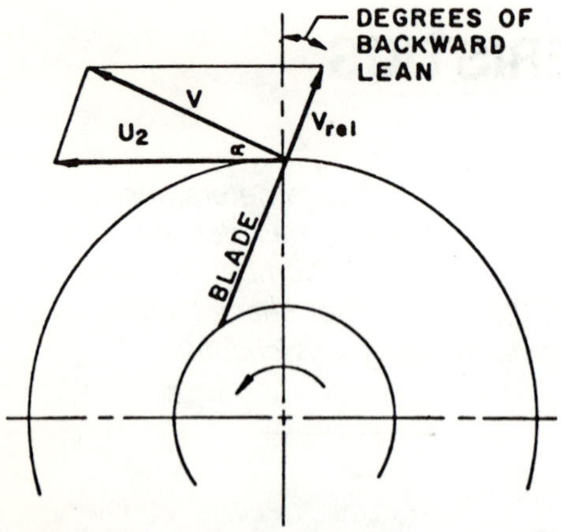

Figure 4.3. Vector diagram of the gas velocity relative to the impeller blade. The slope of the characteristic curve is strongly influenced by this relationship [17]. *(Used with permission of Elliott Company, Jeannette, PA.)*

to the centrifugal compressor performance curve (Figure 4.5).

Figure 4.6 shows characteristic curves for three basic configurations: forward-leaning, radial, and backward-leaning blade profiles. Note that the forward-leaning blades provide a positive sloping head curve and the maximum head output. This is because V_T is increasing with increasing flow.

A radial bladed impeller has a theoretical constant (flat) head curve since V_T does not change with flow.

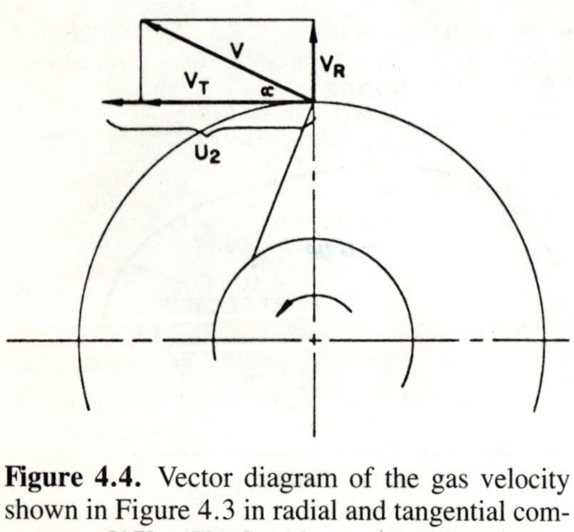

Figure 4.4. Vector diagram of the gas velocity shown in Figure 4.3 in radial and tangential components [17]. *(Used with permission of Elliott Company, Jeannette, PA.)*

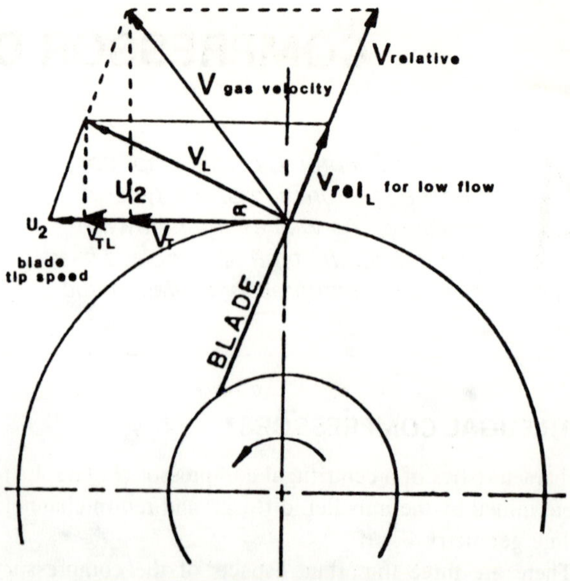

Figure 4.5. The effect of a change in flow rate on the vector diagram at the impeller O.D. is shown. Note that V_T decreases as flow increases (V_{rel} increases) for a backward-leaning impeller blade. This gives the backward-leaning impeller the characteristic negative-sloping head curve shown in Figure 4.2. *(Used with permission of Elliott Company, Jeannette, PA.)*

Overall stage efficiency is highest for backward-leaning impellers, while efficiency is lowest for forward-leaning blades.

For best efficiency, most modern centrifugal compressors use backward-leaning bladed impellers. Directionally speaking, the greater the backward lean, the better the efficiency. However, as the angle increases, the head is reduced (see Figures 4.6C and 4.7). A designer can select a blade angle and tip width to best fit the desired head and efficiency characteristics of the particular application.

STONEWALL

Stonewall, or choke, is a condition at which increased capacity (flow) results in an excessive decrease in head (see Figure 4.9). This occurs because the Mach number is approaching 1.0.

Operating at a very high flow rate has very negative effects on the performance of a centrifugal compressor, and can sometimes be damaging. For axial compressors, high flow rates can also create blade flutter and result in serious damage to the blading. The stonewall effect of the centrifugal compressor stage with a vaneless diffuser is controlled by impeller inlet vector geometry.

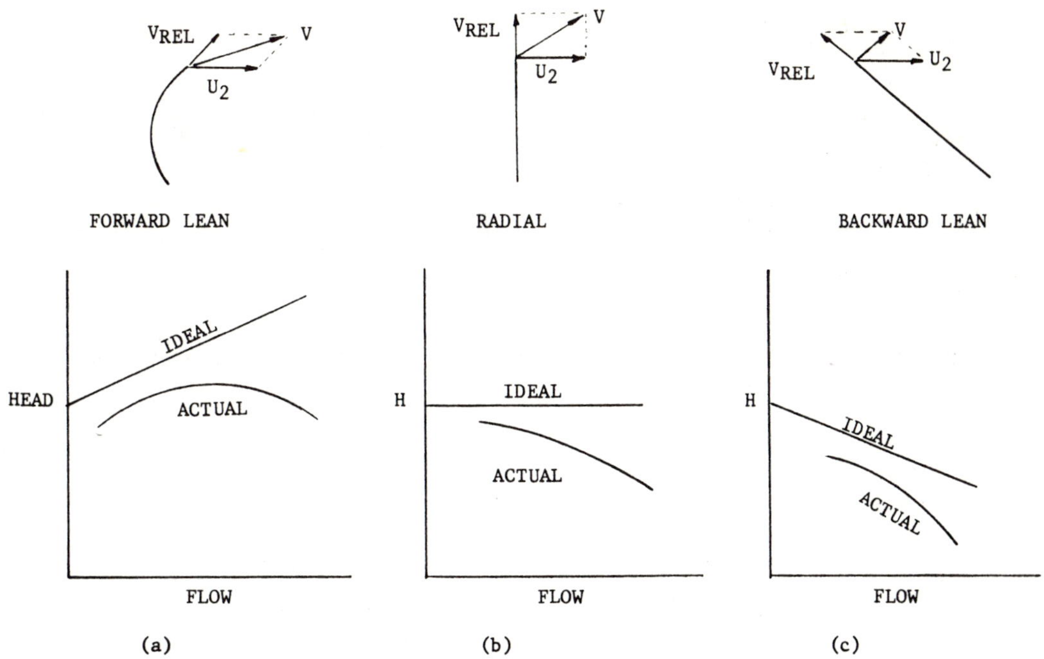

Figure 4.6. Three basic head curve shapes for centrifugal compressors. *(Adapted from [9], p. 67.)*

U_1 in Figure 4.8 represents the tangential velocity of the leading edge of the blade. V represents the absolute velocity of the inlet gas, which, having made a 90° turn is now moving essentially radially (in the absence of prewhirl vanes)—hence the name radial inlet. By vector analysis V_{rel},

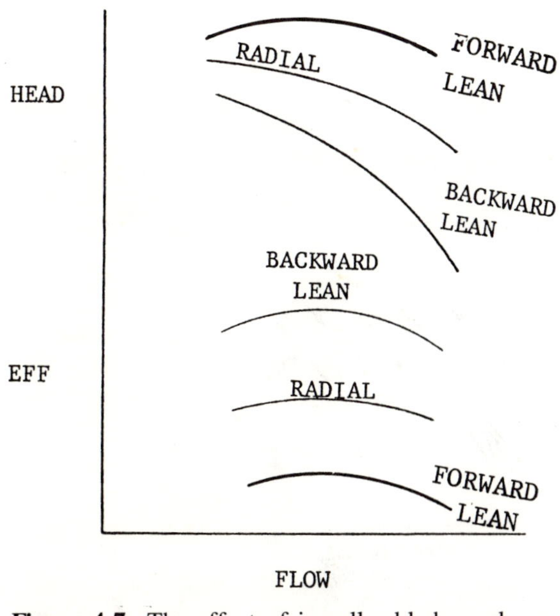

Figure 4.7. The effect of impeller blade angle on head.

which is gas velocity relative to the blade, is of the magnitude and direction shown.

$$V = U_1 + V_{rel}$$

At design flow, V_{rel} lines up with the blade angles. As flow increases beyond design, V increases. As V increases so does V_{rel}. V_{rel} now impinges at a negative angle to the blade, a condition known as negative angle of attack. High negative angles of attack contribute to the stonewall phenomenon because of boundary-layer separation and a reduction of effective area in the blade pack. This area reduction, in addition to the already high V_{rel}, brings on Mach 1 and a corresponding shock wave, as shown in Figure 4.9.

SURGE

Surge flow has been defined as peak head [7]. Below the surge point, head decreases with a decrease in flow (Figure 4.2).

Surge is especially damaging to a compressor and must be avoided. During surge, flow reversal occurs resulting in reverse bending on nearly all compressor components. The higher the pressure or energy level, the more damaging the surge forces will be.

As flow is reduced at constant speed, the magnitude of V_{rel} decreases proportionally, causing the flow angle to decrease. (See Figures 4.5 and 4.10.) Additionally, the incidence angle is increased (Figure 4.11).

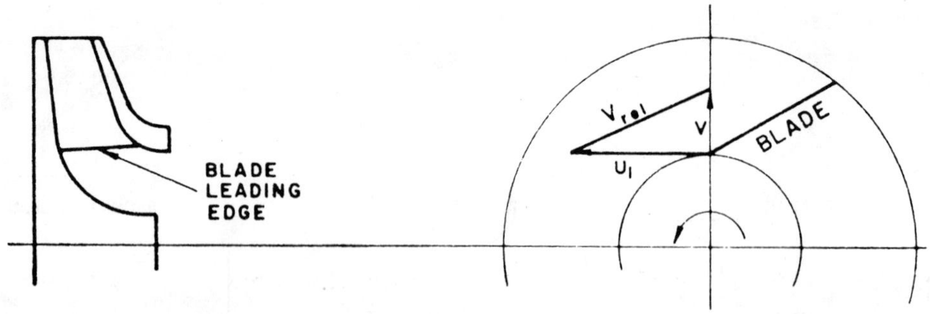

Figure 4.8. Stonewall [17]. *(Used with permission of Elliott Company, Jeannette, PA.)*

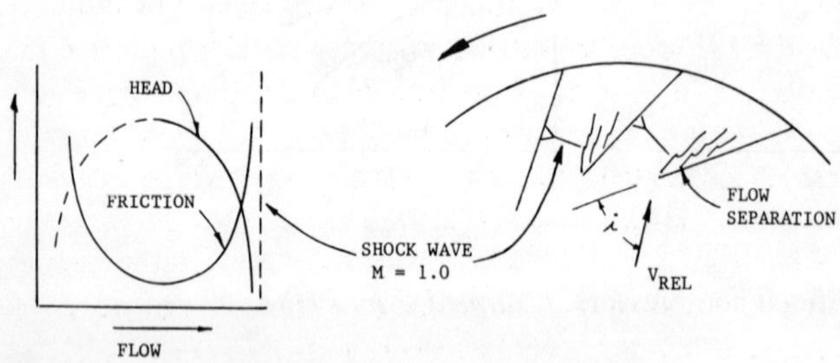

Figure 4.9. Stonewall.

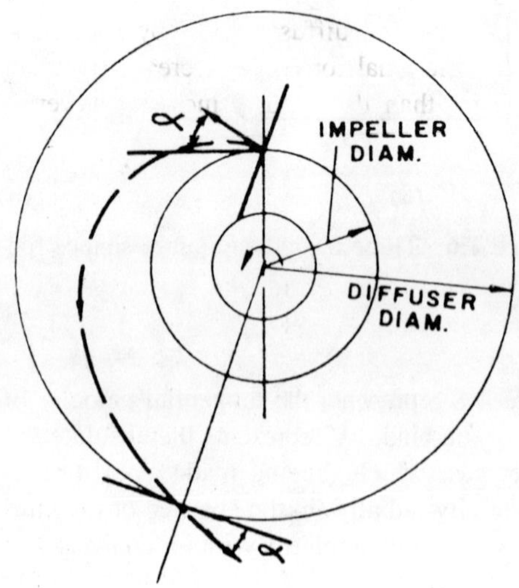

Figure 4.10. Flow through a diffuser [17]. *(Used with permission of Elliott Company, Jeannette, PA.)*

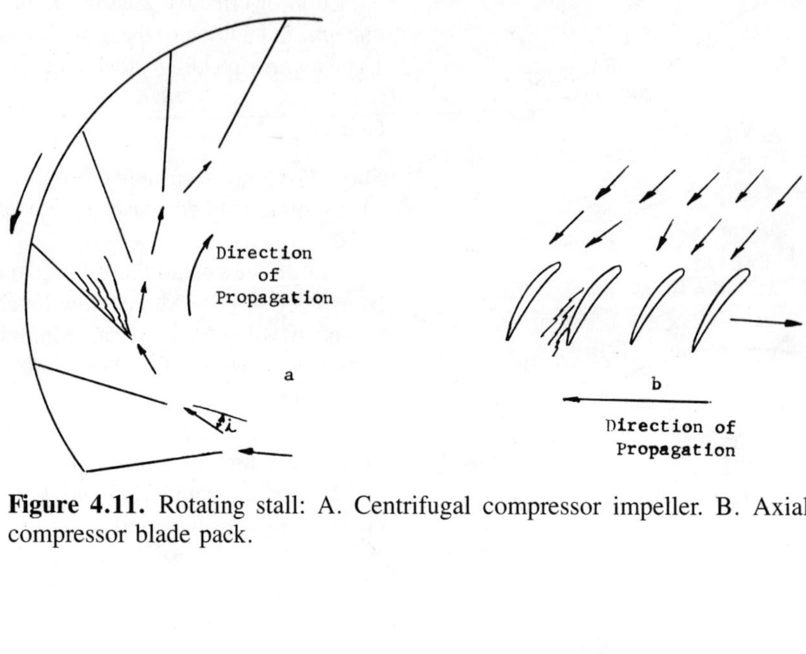

Figure 4.11. Rotating stall: A. Centrifugal compressor impeller. B. Axial compressor blade pack.

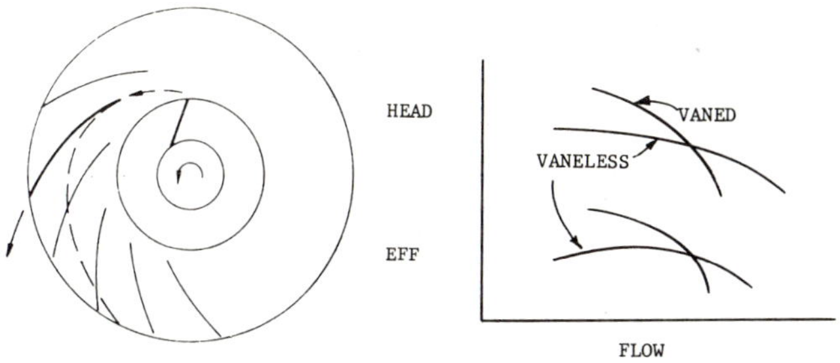

Figure 4.12. Diffuser vanes. *(Data from [10, 17].)*

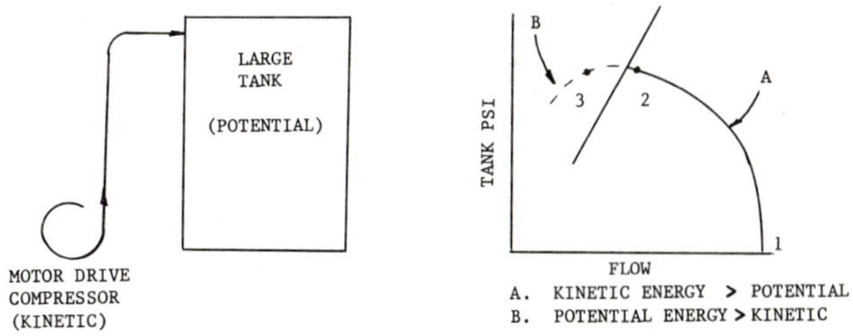

Figure 4.13. Surge.

The smaller the flow angle α, the longer the flow path of a given gas particle from the impeller tip to the diffuser outside diameter. When angle α becomes small enough, and the diffuser flow path long enough, the flow momentum of the gas is dissipated by the diffuser walls by friction to the point where the frictional forces are increasing (versus flow reduction) faster than the head is increasing (versus decreasing flow).

The high losses associated with low flow (see Figure 4.11) are partly caused by a poor incidence angle i, which can result in flow separation at the low pressure side of the blade leading edge. This flow separation frequently starts at one or more blades and continuously shifts around the impeller blades (or blade row in an axial compressor). This occurs at relatively low speeds just before full surge occurs. At higher speeds, the compressor generally goes directly from stable operation to flow separation on all blades and full reverse flow.

The flow separation plus the higher frictional losses result in a positively sloped curve. Since system resistance curves are also positively sloped, the system is unstable.

The point at which a compressor surges can be controlled somewhat by the designer's adjusting the diffuser area to increase V_R and flow angle α. Of course, higher velocities result in higher frictional losses, so a designer must balance between desired surge point and stage efficiency during the design process.

The surge point is reduced by the addition of vanes in the diffuser (Figure 4.12). The vanes shorten the flow path through the diffuser, reducing frictional losses and controlling the radial velocity component of the gas. Due to lower friction, head and efficiency are enhanced, but the operating range is reduced. Off-design operation rapidly changes the incidence angle to the vanes and flow separation occurs, resulting in the reduced operating range.

To better understand what is occurring during surge, visualize the simple system shown in Figure 4.13. The system consists of a small motor-driven compressor delivering air to a relatively large tank. While in an idle state, the entire system is at ambient conditions. The instant the unit reaches design speed, the pressure in the tank is still zero (Point 1). As time passes, the pressure builds in the tank and flow is reduced due to increased resistance. Eventually Point 2 is reached where the pressure of the tank causes such a high backpressure on the compressor that flow through the impeller is significantly reduced. Much of the energy input is going to friction instead of building head. This is due to both the mismatch of inlet angle and the longer diffuser passage described earlier. Since this effect continues to build as flow is reduced, the slope of the head curve is reversed. As flow is reduced to Point 3, the head output of the compressor is also reduced. Since the pressure in the tank is still at Point 2, flow occurs from the tank to the compressor. Once the pressure in the tank is reduced (by reverse flow) to a level less than the head capability of the compressor, the process will then recover and the gas will flow from the compressor to the tank. This process will continue to repeat itself indefinitely.

OFF-DESIGN OPERATION*

Off-design operation of a compressor can dramatically affect the performance characteristic curve shape. Any change in inlet conditions can change the discharge pressure and gas horsepower as shown in Figure 4.14. (See also Conceptualizing Head, Chapter 2.)

In addition to changing the characteristic pressure and horsepower curves the characteristic head curve also changes. This is due to volume ratio effects and equivalent speed effects.

If a constant discharge pressure is desired and gas conditions are changed (inlet pressure or temperature, mole weight change), a speed change is required. Since the curve shape changes with speed (higher losses at higher speeds), the head curve shape then changes (Figure 4.15). This effect is further compounded by volume ratio effects (Figure 4.16).

The head characteristics are a function of the acoustic velocity of the gas (see Figure 4.9). Knowing this, it is most convenient to refer to some "constant" gas, and obtain an "equivalent tip speed." This reference constant is typically

*Adapted from "Centrifugal Compressors . . . The Cause of the Curve," D.C. Hallock, Elliott Company, Jeannette, PA, 1968, with permission [17].

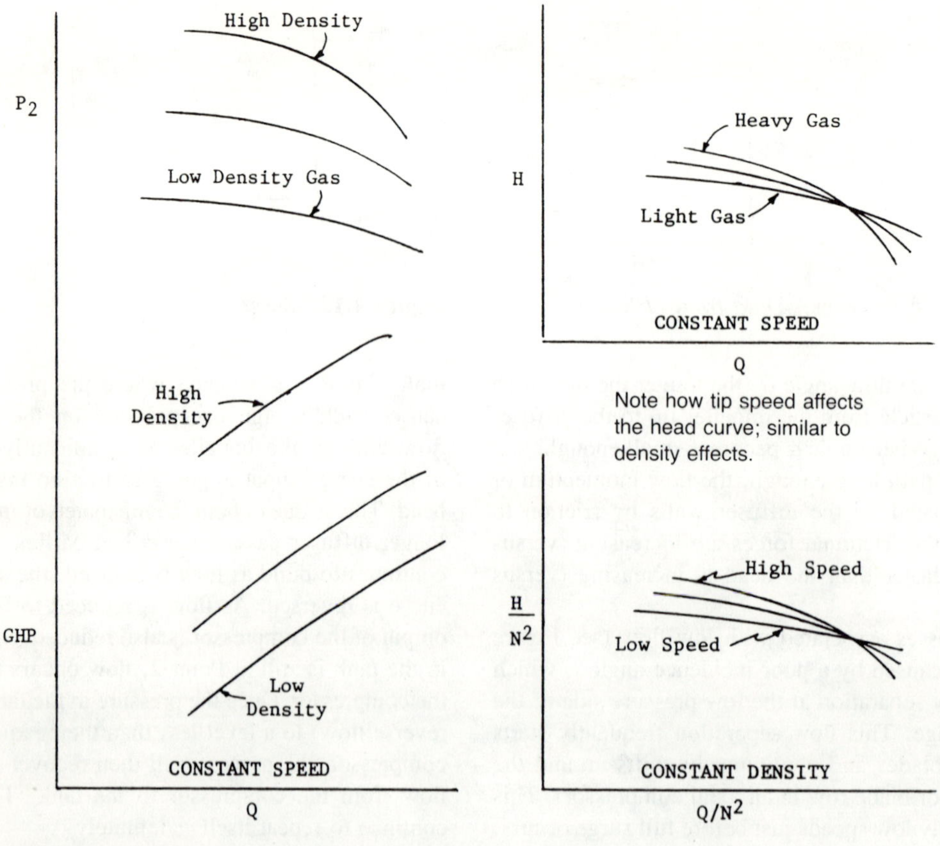

Figure 4.14. Effect of varying inlet conditions at constant speed for a single-stage compressor. For a multi-stage compressor, the curve shape and operating range is further compounded by volume ratio effects. See Figure 4.16. *(Data from [11, 17].)*

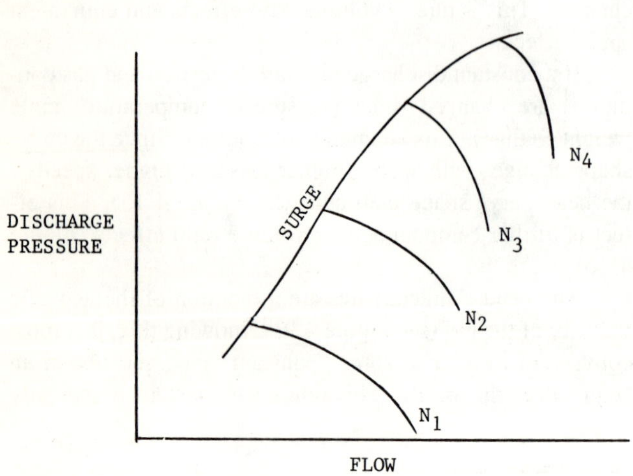

Figure 4.15. Effect of speed change on compressor curve shape.

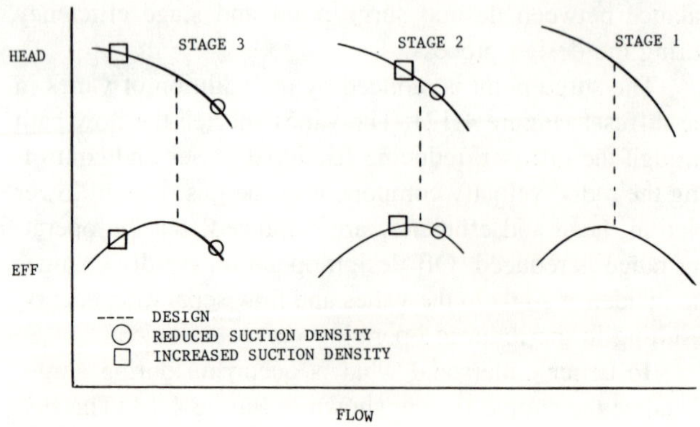

Figure 4.16. Volume ratio effects. *(Data from [11].)*

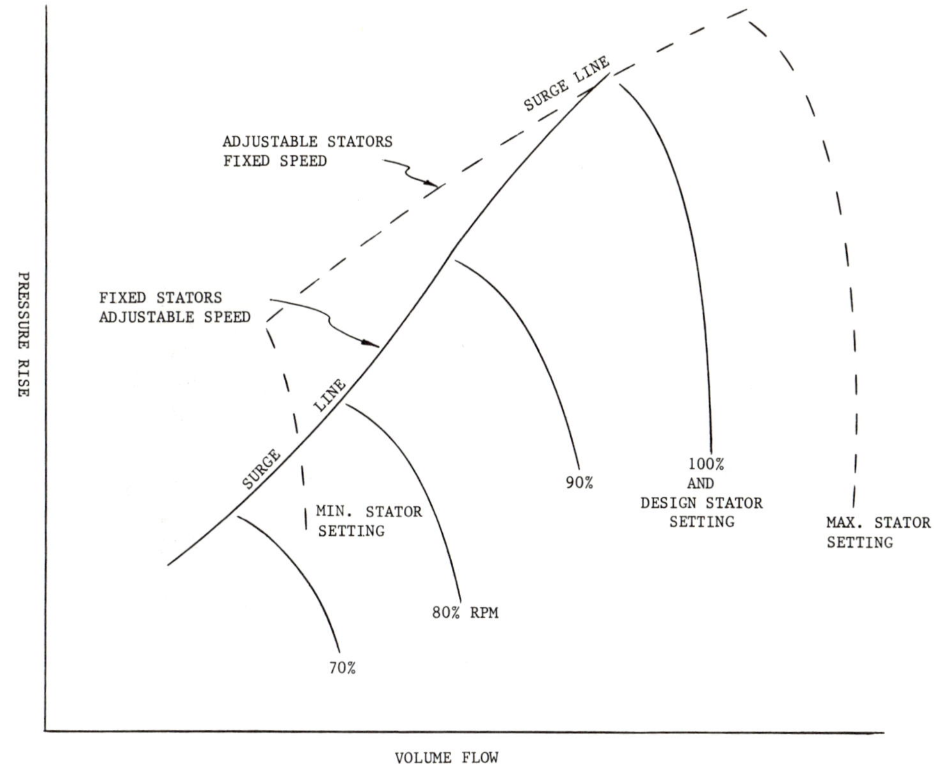

Figure 4.17. Axial compressor performance map showing effect of adjustable stator vanes.

air at 80°F, since this is what most "developmental" testing uses as a test medium.

As an example, we know the sonic velocity of air at 80°F is 1140 fps, and that of propylene at −40°F is 740 fps. If a compressor stage is operating at a mechanical tip speed of 780 fps on propylene at −40°F, the equivalent tip speed is

$$\begin{aligned} U_{eq} &= 780 \times \frac{1140}{740} \\ &= 1200 \text{ fps} \end{aligned}$$

The stage characteristic head curve shape at 780 fps on propylene is therefore the same as 80°F air at 1200 fps. (Note: This is not exact. There is also an impeller tip volume ratio effect based on gas density that causes head and work input to change somewhat. An air and propylene test will not result in exactly the same head curve at constant equivalent speed, but for all practical purposes the results are close enough to be considered the same.)

In a multi-stage compressor the "equivalent speed" effect is compounded by volume ratio effect (Figure 4.16). If the gas density varies, the pressure rise and volume ratio will also vary. This will feed a different flow rate to the second stage. The effect on following stages will be compounded. The end result is a premature choke and surge.

ADJUSTABLE VANES

Adjustable vanes can be used to extend the useful operating range of any compressor (see Figures 4.17 and 4.18). Adjustable vanes are typically used on the first several stages of axial compressors since they are very sensitive to the gas angle of attack and have a very short operating range. Adjustable vanes are also very popular for single-stage centrifugal units and are occasionally used on multi-stage centrifugal compressors.

The range of operation is extended by changing the angle of attack on the inlet side of the impeller blade (Figures 4.1 and 4.8). For the high flow region, the angle of attack can be enhanced to eliminate flow separation and

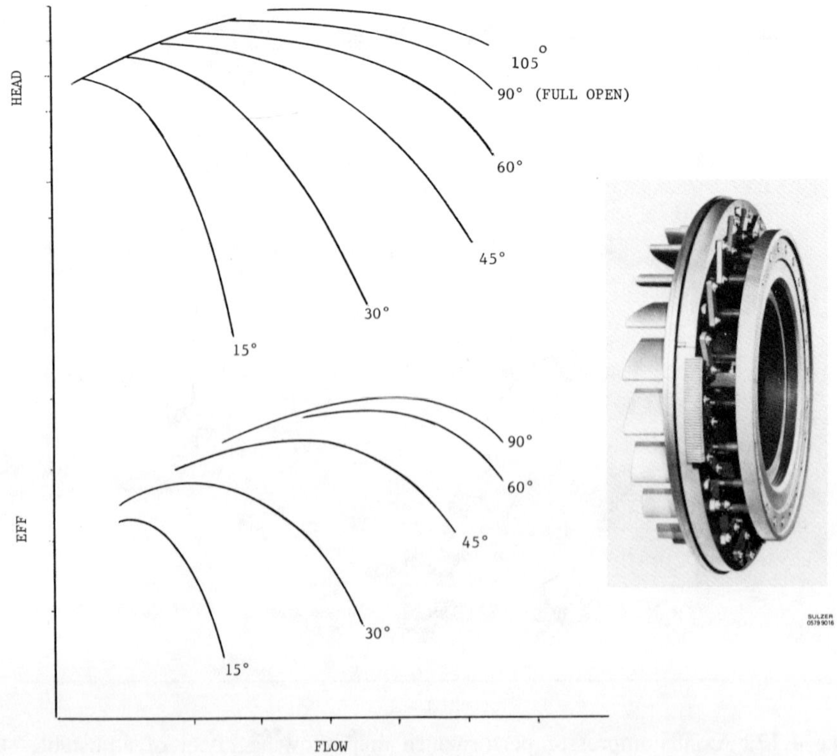

Figure 4.18. Adjustable inlet guide vanes. Curve shows effect of various vane positions on head and efficiency for single-stage centrifugal compressor. Photo shows inlet guide vanes used on the first stage of a multi-stage centrifugal compressor. *(Photo courtesy of Sultzer-Escher Wyss Ltd.)*

effectively increase the "throat" area of the impeller. This will increase the capacity of the impeller. Also, head will increase due to "against" rotation swirl (see Figure 4.9). By adjusting the vanes to provide swirl in the direction of the impeller rotation, V_{rel} is reduced, reducing head (Figure 4.18). Since the incidence angle is improved, frictional losses are improved and peak efficiency as well as peak head shifts toward reduced flow.

AXIAL COMPRESSORS*

The operating principals of the axial compressor differ significantly from those of the centrifugal in that the compressor characteristics are dependent on the lift and drag coefficients of the cascade of airfoil blades (Figure 4.19). The nominal axial velocity of the gas is constant throughout the axial, while in the centrifugal compressor the gas is being accelerated and decelerated. The reduced wetted perimeter that the gas "sees" and the short flow path through the axial contribute to the improved overall efficiency.

Figure 4.20 shows general lift and drag characteristics of an airfoil. This is fine for an airplane wing but for a compressor this information must be in the form of head and inlet flow as in Figure 4.21.

Because relatively low pressure rise is available from a single axial stage, relatively high speeds are utilized to maximize its effectiveness. Mach numbers are relatively high, so any change in inlet conditions, such as suction temperature, can dramatically affect the performance curve (Figure 4.22).

Since Mach number, $M = V/a$, and sonic velocity, $a = \sqrt{kgRT}$, it is clear that for a given gas, Mach number is directly related to the temperature of the gas.

*Based on "Fundamentals of Fluid Flow as Applied to the Design of Axial Flow Compressors and Fans," W.K. Bodger, Carrier Corporation, 1954, with permission [1]; and "Axial Compressor Design Philosophy," P. Whiteman, Elliott Company, Jeannette, PA, with permission [19].

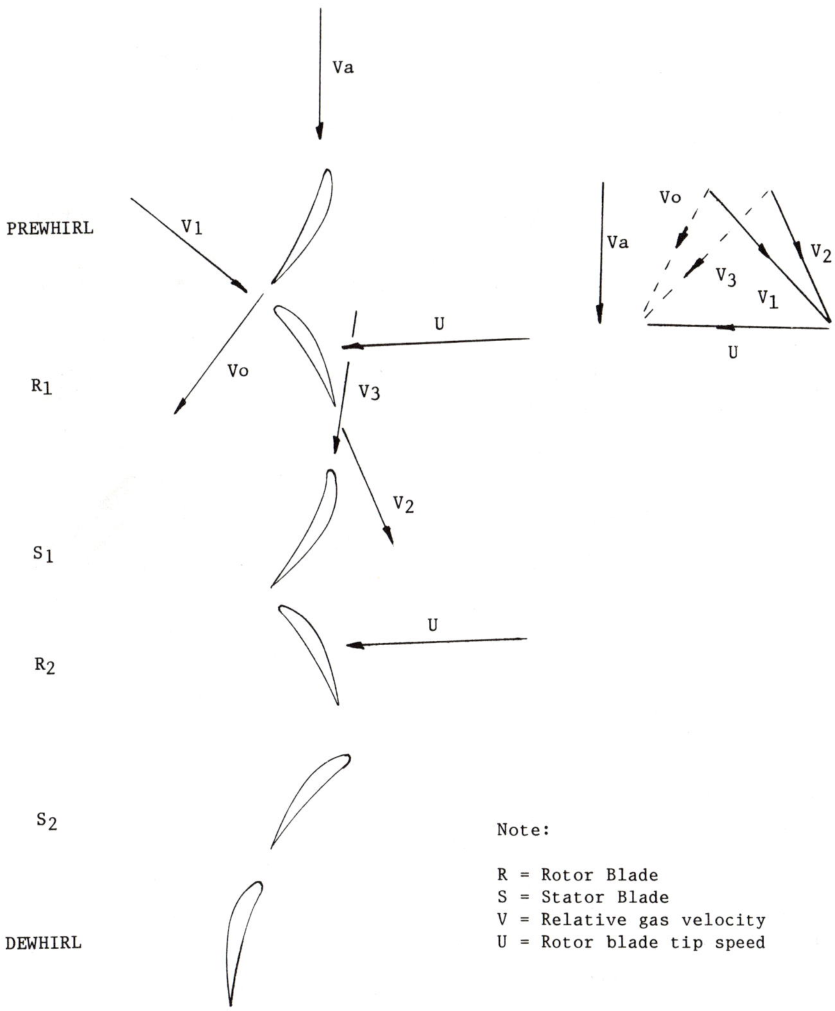

Figure 4.19. Axial compressor vector diagram.

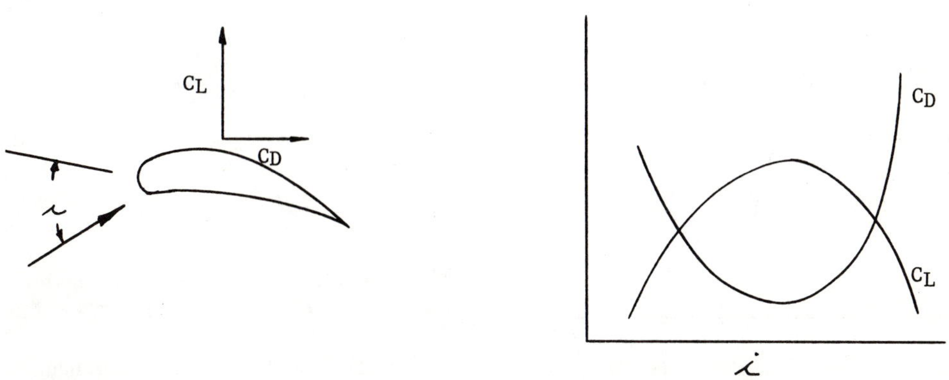

Figure 4.20. Axial compressor staging is based on airfoil lift and drag coefficients. *(Data from [1].)*

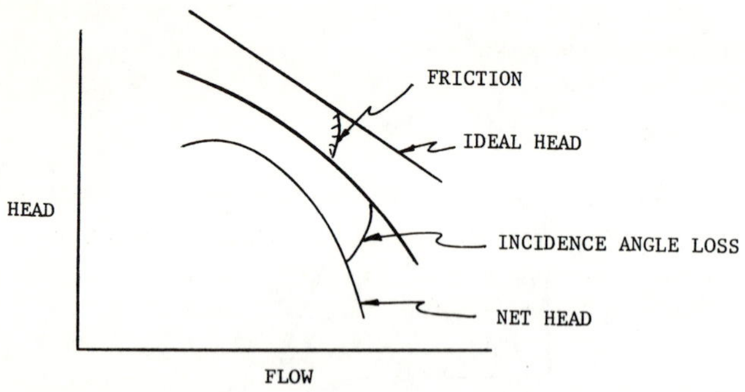

Figure 4.21. Losses in an axial compressor stage are due to friction and incidence angle loss. *(Data from [1].)*

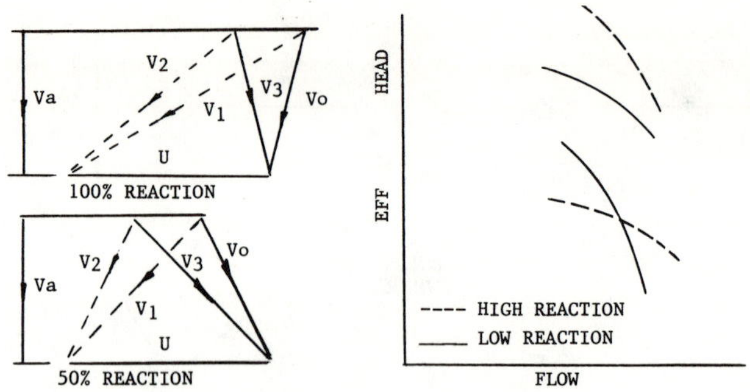

Figure 4.23. Reaction. (*Data from [19].*)

REACTION

Reaction is the degree of pressure rise in the rotating vanes versus overall pressure rise of the stage. Fifty percent reaction is the most popular, as this gives the best overall efficiency. In a 50% reaction stage, an equal amount of diffusion—or pressure rise—takes place across the rotating and stationary blades. In a 100% reaction stage, the total pressure rise is in the rotating blade while the stator simply acts as a turning vane (Figure 4.23).

SURGE

From Figure 4.20 it can be seen that a low angle of attack (low flow) results in high drag (low efficiency) and a low coefficient of lift (low head). As with the centrifugal compressor, the positive-sloped portion of the curve is unusable since this area of operation gives an unstable system and reverse flow occurs. Surge is especially damaging to an axial compressor because of the relatively large mass flow rate of gas and relatively thin blades. Besides the problem of reverse bending stresses and eventual fatigue, there is a problem of thermal growth. During surge, discharge gas

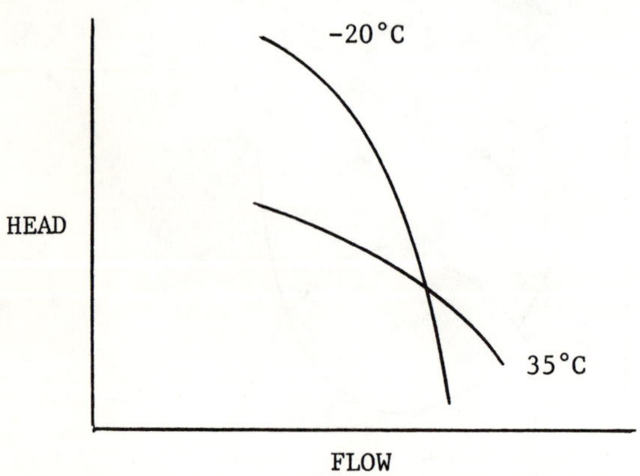

Figure 4.22. Effect of suction temperature change on axial performance curve. *(Data from [19].)*

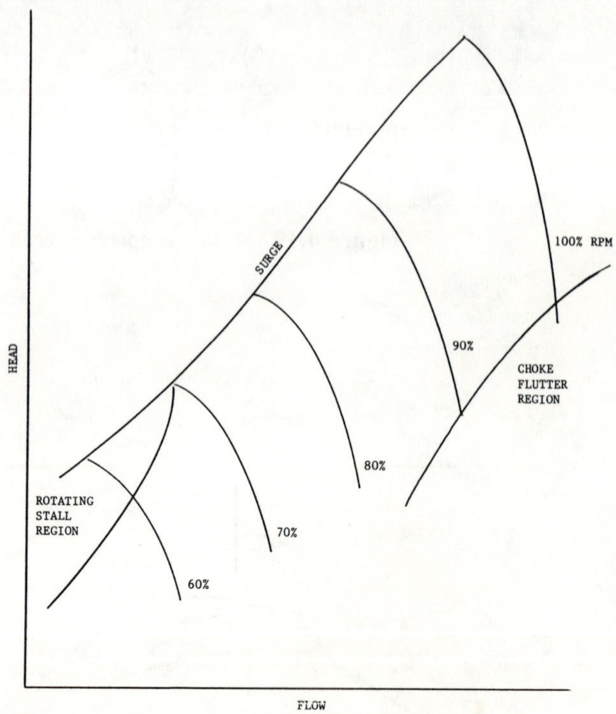

Figure 4.24. Axial compressor performance map showing the choke flutter region. It is not easy to determine when choke flutter is occurring. Mounting vibration or strain gauge equipment on the stator vanes may be necessary to determine the presence of choke flutter.

is being forced back through the compressor then recompressed. The compressor is now compressing heated gas and temperatures rise quickly, causing the blades to grow, eventually resulting in a rub.

At reduced speeds (70% of design or lower), rotating stall is almost unavoidable (Figure 4.24). However, since rotating stall is only localized and temporary, the overall system is stable, and steady through-flow occurs keeping the blades cool. Since energy levels are lower, blade fatigue is not a serious concern. Even so, it is wise to minimize operation at this point.

CHOKE FLUTTER

At high flow rates the incidence angle becomes reversed and will eventually cause flow separation. This condition is called choke flutter and can be very damaging to the compressor blades (Figure 4.24).

II
APPLICATION

5

EQUIPMENT SELECTION

Whether purchasing a new piece of equipment or rerating an existing unit, it is important to know at least some basic design limits. You can rough out the preliminaries yourself before going to the equipment manufacturer.

When purchasing equipment, define more than a single operating point. Look at your process. Know it well before specifying anything to the compressor manufacturer. Know the type of equipment available and what its characteristics are (Figure 5.1; see also Figure 1.7). A properly matched compressor is applied to the full range of expected operation. Sometimes it may be wise to alter the process somewhat to provide a better match to the capabilities of the compressor.

NEW EQUIPMENT SELECTION

New equipment is the easiest to select since you are starting with a clean sheet of paper. It's just a matter of what equipment is available and how much can be spent. Don't forget that the control system and its logic can be more critical than the compressor selection. Remember, consider a "range" of operation, not a single point.

To start, however, select a single operating point. Using Equation (7.4) and knowing the desired inlet and discharge conditions, determine the total head required for the compression equipment. To determine the number of compression stages required, some vendor data will be needed. Table 5.1 is an overview of what to expect. Note the higher head capability of the open impellers vs. the covered impellers. The cover increases the impeller stress for a given speed. Thus, the open impellers can operate at higher speeds and therefore provide higher head (Figure 5.2). Since open impellers require close clearance to the upstream inlet piece or diaphragm (Figure 8.6), use of open impellers is usually limited to single-stage compressors or to the first one or two stages of a multi-stage centrifugal compressor. Open impellers have also been used as the last stage in a multi-stage axial compressor.

All equipment that you select, regardless of the manufacturer, will have to be matched to existing available equipment, whether off-the-shelf or custom-built. Even with custom-built equipment, the manufacturer generally designs a compressor using a building-block approach of existing components. It is rare, and expensive, to build a new piece of equipment from scratch for a given application.

Custom-design equipment is usually made up of several "families" of compression stages. There could be one family for high head, and another for high efficiency, and a middle-of-the-road family group of compression stages. These families are then scaled to various frame sizes to offer a wide selection, making it possible to customize for a given need by custom selection of standard components.

Although the families generally consist of 20 to 30 impeller sizes for a given frame size, this can be increased to an infinite number to enhance selectability. On one hand, it is important to keep this number to a minimum in order to minimize the number of drawings, castings, stampings and performance records. Although this is a good idea, it is possible with the computerized equipment available for analysis, drafting, and manufacturing to have an infinite number of compression stages per compression family.

Although manufacturers' frame sizes are well-established standards, it is feasible that an infinite number of frame sizes could be made available with use of fabricated components, FEM analysis, and CAD. What this means is that a compression stage can be selected exactly at its peak efficiency instead of compromising and selecting the nearest frame size and the nearest impeller size (Figure 5.3).

There is still the drawback that, while the compressor may be very efficient at the design point, efficiency drops when operating at other than design conditions. To compensate for this when purchasing a new compressor, specify a *range* of operation. Note the full operating range

TABLE 5.1 Approximate per Stage Head Capability of Various Compression Elements

Stage	Head ($Ft\text{-}Lb_f/Lb_m$)	r_p for Air
Centrifugal Compressor		
Covered Wheel	8000–12000	1.3–1.5
Open Wheel	30000–60000	2.6–8.9
Mixed Flow		
Covered Wheel	6000–8000	1.2–1.3
Open Wheel	20000–45000	1.9–3.8
Axial Compressor	4000–5000	1.15–1.20

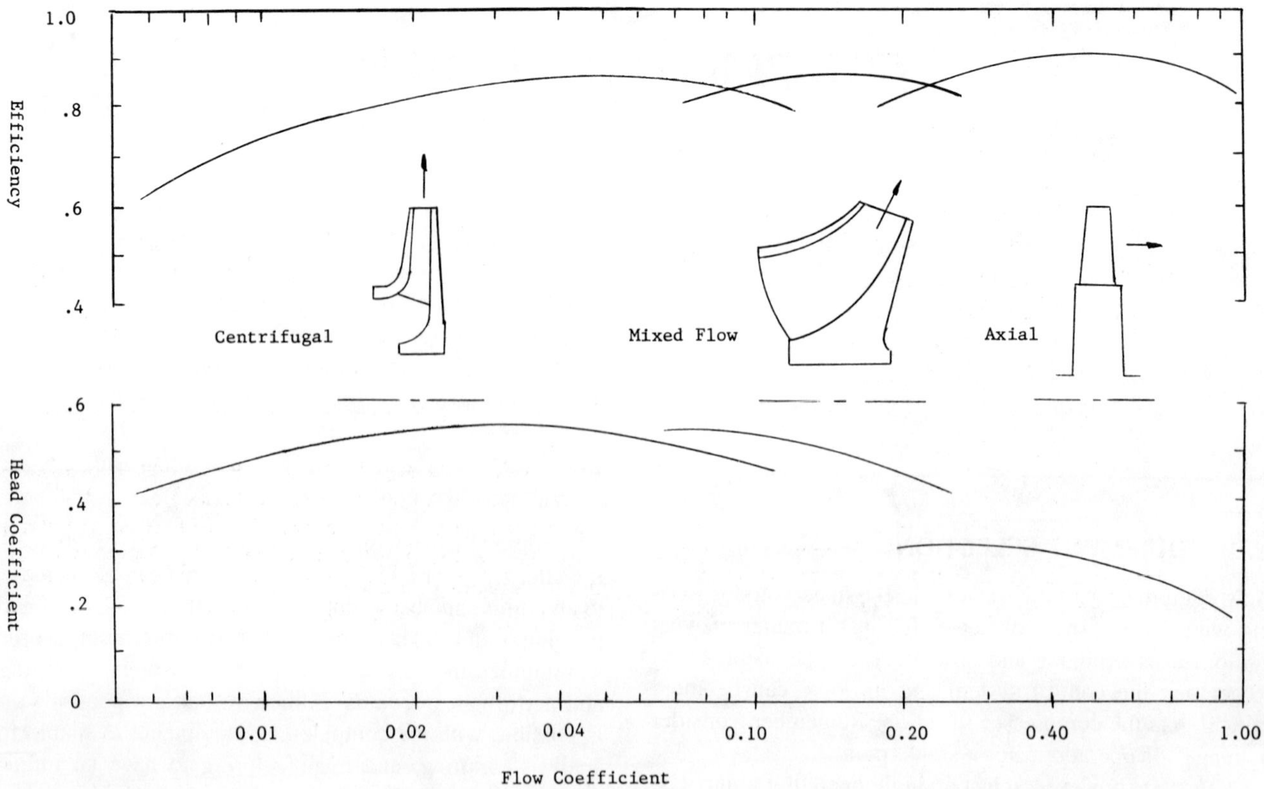

Figure 5.1. Head and pressure capabilities for various compression elements.

of flow, pressures, temperatures, and MW that you expect. Although some process conditions may call for continuous operation at one point, it is common to see some variation in operating conditions.

Consider startup and shutdown conditions. The entire operating range plus startup and shutdown is important for mechanical (rotor dynamics, seals, auxiliary systems, etc.) as well as aerodynamic considerations. Include any purge gas that may be used on startup (nitrogen, carbon dioxide, air, etc.).

Generally speaking, for a given flow, the smaller the frame size the better. The smaller size means a lower price, and usually a better efficiency. Look at Figure 1.6. From this it can be seen that there is an optimum flow for each compressor style. Figure 5.1 shows this even more clearly. The smaller size compressor requires a higher flow coefficient. For even higher flow coefficient compression elements, a mixed flow (Figure 5.4) or an axial compressor is used.

For a single-stage compressor it is wise to choose a mid-sized (flow coefficient, Equation (2.60)) stage to optimize efficiency. For multi-stage compressors, the first stage must be a high-flow stage so that use of very low-flow stages can be limited. Additionally, double-flow compressors can be used as the first section in a compressor, or as the first body in a string of equipment. Mechanical limitations must also be considered, such as maximum tip speed (stress limits), and stage spacing (critical speeds). High flow and high efficiency generally mean larger stage spacing (axial length), while high head and low flow generally mean reduced stage spacing. High flow and high efficiency design staging may require reduced speeds due to higher stress levels encountered. Corrosive elements in the gas can also be a cause of limited speed. H_2S requires that controlled yield (80,000 to 90,000 psi) material be used for the impellers, thus limiting the operating speed.

For a given flow coefficient, larger equipment frame size means better efficiency. The larger size results in an improved "wetted perimeter" and thus reduced friction similar to using a larger pipe for fluid flow. It is better to use one large compressor rather than two small compressors in parallel. But the effect of stage size (flow coefficient, Figure 5.1) may outweigh the frame-factor effect (Table 5.2). For example, assume an application where a compression stage for 60,000 CFM is required. Frame E in Table 5.2 would meet this need. However, for improved mechanical

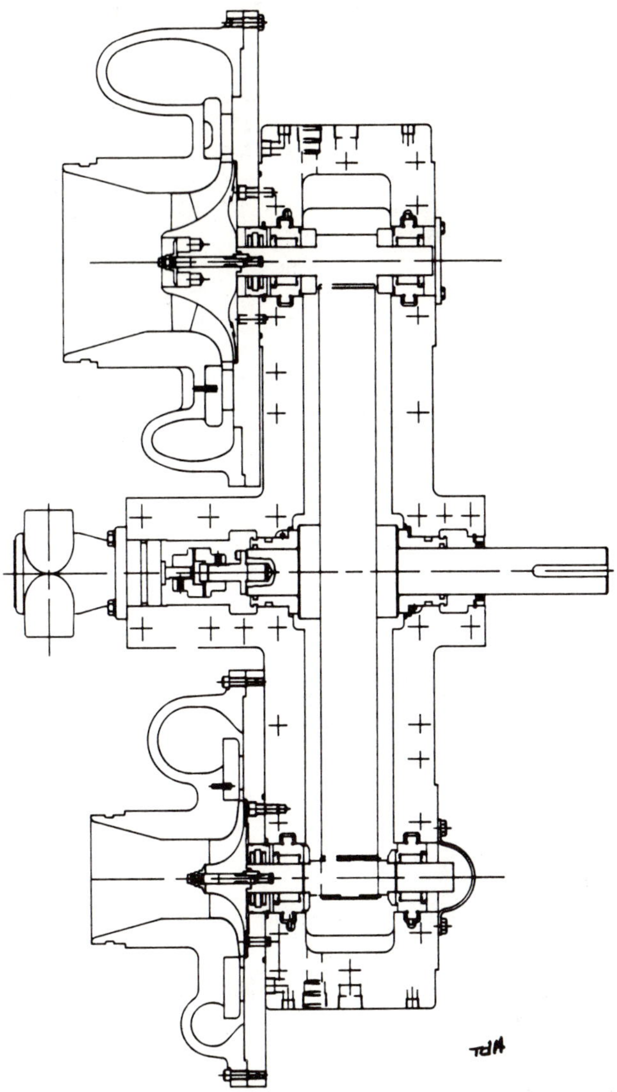

Figure 5.2. High-speed, open wheel centrifugal compressor. *(Used with permission of Elliott Company, Jeannette, PA.)*

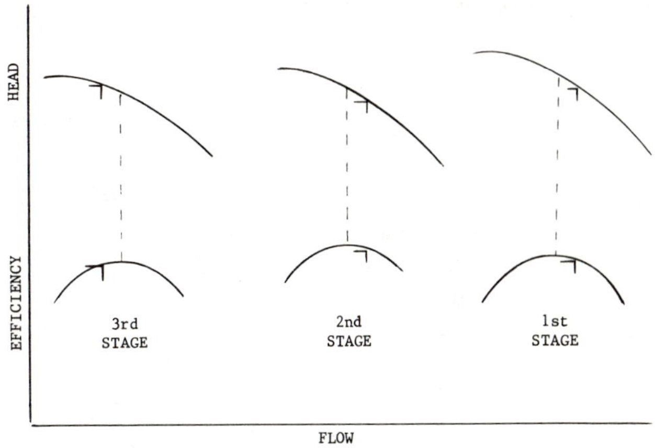

Figure 5.3. Selection of a three-stage compressor utilizing a "fixed" family of compression stages. Note that in each case the design point is removed from the peak efficiency point. An infinite number of stages to select from would improve overall efficiency by allowing the design point to always be at maximum efficiency.

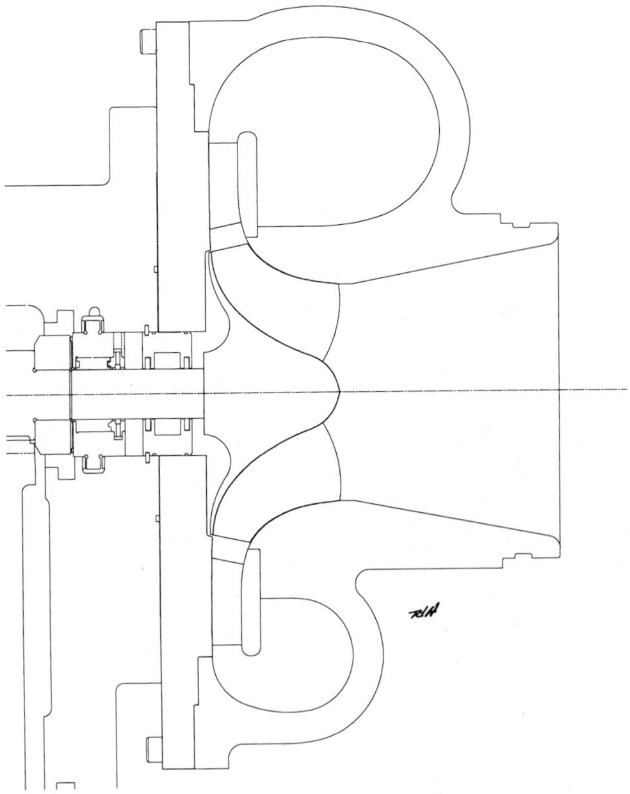

Figure 5.4. Mixed-flow compressor. The mid-flow region (see also Figure 5.1) is enhanced by the use of hybrid designs called mixed-flow stages which use three-dimensional blade designs. In addition to improved efficiency, the mixed-flow compressor can result in a 30 to 40% smaller package (volume and weight) for a given flow.

TABLE 5.2 Typical Centrifugal Compressor Frame Data

Frame	Nominal Flow Range (cfm)	Maximum Number Stages	Nominal Speed (rpm)	Nominal Polytropic Efficiency	Nominal H/N^2 Per Stage	Maximum Q/N
A	250–2500	10	15000	.74	3.7×10^{-5}	0.15
B	800–9000	9	11500	.76	7.5×10^{-5}	0.65
C	5000–25000	9	8000	.77	1.5×10^{-4}	2.9
D	15000–35000	8	6000	.78	2.5×10^{-4}	5.5
E	30000–70000	8	5000	.78	4.0×10^{-4}	12.0
F	55000–125000	8	3000	.79	10×10^{-4}	50.0
G	100000–170000	8	2700	.79	12×10^{-4}	65.0

reliability, a better selection might be two D frames to operate in parallel. The single E frame would require less power if the same (scaled) impellers are used for both the D and E frame selections.

SELECTION PROCEDURE

1. Start by determining the gas mixture properties. This can be done by referring to Tables 2.1 and A.1 (in Appendix A). Additionally, refer to Example 7.4. If a Mollier diagram is available, use it.
2. Next, calculate the actual inlet volume flow for the conditions at the compressor suction flange. Normally expressed in cubic feet per minute, this flow is designated as ACFM (actual cubic feet per minute) or ICFM (inlet cubic feet per minute). Use the gas properties from Step 1 to make the conversion required.
3. Select the compressor frame size using Table 5.2.
4. Calculate total head required. See Equation (7.4).
5. Calculate the total number of stages required. Refer to the vendor literature on nominal head or H/N^2 values as shown in Table 5.2. Multiply this value by the nominal speed squared. Divide the total head by the per-stage head to approximate the number of stages.
6. Adjust the speed to obtain the discharge conditions desired by using fan laws, Equations (2.66) and (2.67).
7. Gas horsepower should be corrected for balance piston leakage and mechanical losses, which should be available from vendor literature.

An outline of items for consideration during the selection process is provided in Table 5.3.

TABLE 5.3 Selection Outline

Rough out equipment yourself to get a "feel" for what you are buying.

Specify a range of operation.

If applicable, specify upratability for capacity and pressure.

Consider shop and follow-up field performance testing.

Provide for proper inlet piping.

Consider abradable seals and diffuser coatings to maintain high efficiency.

Review also the following:

- Driver power
- Foundation
- Torque (shaft stress)
- Lateral and torsional criticals
- Couplings
- Lube system
- Casing pressure ratings
- Impeller stress levels
- Nozzle and volute velocities
- At-speed balancing of the rotor

Provide for off-design conditions:

- Air dryout
- Nitrogen purge
- Start-up
- Shut-down
- Peak load
- Off load
- Standby
- Other

EXAMPLE 5.1
N-Method

1. Given the following conditions,

$\dot{M}$ = 1769 lb/min   MW = 29
P_1 = 80 psia k = 1.4
T_1 = 90°F (550°R) Z = 1.0
P_2 = 225 psia

2. Calculate inlet volume.

$$v_1 = \frac{ZRT_1}{144P_1} = \frac{1.0(1545)(550)}{144(29)(80)} = 2.544 \text{ ft}^3/\text{lb} \quad (2.7)$$

$$Q = \dot{M} \times v_1 = 1769 \times 2.544 = 4500 \text{ ICFM}$$

3. Select compressor frame size.
   Based on an inlet volume of 4500 ICFM and knowing the required discharge pressure is 225 psia, select a B frame size from Table 5.2.
4. Calculate the required head.
   Assume an efficiency of 0.76 from Table 5.2 and calculate the polytropic exponent.

$$\frac{n}{n-1} = \frac{k}{k-1}\eta_p = \frac{1.4}{0.4}(0.76) = 2.66$$

Calculate the overall head.

$$H = ZRT\left(\frac{n}{n-1}\right)\left[\left(\frac{P_2}{P_1}\right)^{(n-1)/n} - 1\right]$$

$$= 1.0\left(\frac{1545}{29}\right)(550)(2.66)\left[\left(\frac{225}{80}\right)^{0.3759} - 1\right]$$

$$= 37029 \frac{\text{ft-lb}_f}{\text{lb}_m}$$

Check the discharge temperature for a need to intercool (cool if $T_2 > 400$°F).

$$\frac{T_2}{T_1} = \left(\frac{P_2}{P_1}\right)^{(n-1)/n} = \left(\frac{225}{80}\right)^{0.3759} = 1.475$$

$$T_2 = 550(1.475) = 811°\text{R} = 351°\text{F}$$

Iso-cooling is therefore not required.

5. Determine the number of casing stages.
   From Table 5.2 the nominal speed for a B frame is 11,500 RPM. Calculate the Q/N.

$$Q/N = \frac{4500}{11500} = 0.391$$

From Table 5.2, $H/N^2 = 7.5 \times 10^{-5}$
H/stage would then be

$$H/N^2 \times N^2 = (7.5 \times 10^{-5})(11500)^2 = 9919 \frac{\text{ft-lb}_f}{\text{lb}_m}$$

Determine approximate number of casing stages.

$$\text{Number of stages} = \frac{37029}{9919} = 3.75.$$

Four stages are required.

6. Adjust speed.
   Adjust the nominal speed according to the casing stages. Four stages must develop 37029 ft-lb$_f$/lb$_m$ or an average of 37029/4 = 9257ft-lb$_f$/lb$_m$ per stage.
   Using Fan Law relationships adjust the speed.

$$H \propto N^2$$

$$N = N_{\text{NOM}}\left(\frac{H_{\text{REQ'D}}}{H}\right)^{1/2} = 11500\left(\frac{9257}{9919}\right)^{1/2}$$

$$N = 11{,}110 \text{ RPM}$$

7. Calculate the approximate power.

$$\text{GHP} = \frac{\dot{M} \times H}{33000 \times \eta_p} = \frac{1769 \times 37029}{33000 \times 0.76} = 2612 \text{ HP}$$

Adjust for balance piston leakage.

$$2612 \times 1.02 = 2664 \text{ HP}$$

Add losses for bearings and seals (data from vendor literature).

$$\text{SHP} = 2664 + 65 = 2729 \text{ HP}$$

RERATES*

Frequently, process requirements change due to market fluctuations, new concepts in processing, or other reasons. Because of this, the compressor can be called upon to operate in a range for which it was not originally designed.

One way to satisfy the needs of the new conditions is of course to buy new equipment. A rerate, however, can often provide the means of meeting these new conditions at significantly less cost than new equipment, since much of the hardware can be reused.

A rerate can be difficult in the sense that the Application Engineer is very limited. The casing size and bearing span is fixed. Capacity and pressure rise is therefore limited.

To determine the rerate feasibility of a compressor rerate, the following must be considered:

1. *Capacity*: Are the nozzles large enough to accept the new flow rate?
2. *Horsepower*: Can the motor, turbine, or gear handle the new horsepower?
3. *Pressure*: Can the compressor handle the new pressure rise? Is the casing able to meet the new pressure levels?
4. *Speed*: Can the rotor handle the new speed? What are the effects on rotor dynamics and impeller stress?

CAPACITY

The primary factor in considering the capacity limit is whether the nozzles will pass the required flow at a reasonable pressure drop. A good rule of thumb for inlet nozzles for air or light gas is 140 fps. Otherwise use a velocity such that

$$\frac{P_v}{P_1} \leq .01 \qquad \text{(See Figure 6.17)}$$

where

$$P_v = \frac{P_1 V_1^2}{2gZ_1RT_1}$$

This should be calculated using the area schedule at the inlet flange. For discharge nozzles, velocities can be about 35% higher. The discharge volute can also be the limiting factor. However, only the OEM can properly determine this.

HORSEPOWER

The rebuilt compressor will require a power increase approximately proportional to the increase in weight flow (see Equation (7.12)). This means that an increase in weight flow of 20% will require an increase in horsepower of at least 20%.

Another consideration is pressure rise or head. Horsepower is directly proportional to head.

$$\text{GHP} = \frac{\dot{M}H_p}{33000\eta_p} \qquad (7.12)$$

If an increase of 20% in weight flow is coupled with an increase in polytropic head of 20%, the power will rise by 44%.

$$1.20 \times 1.20 = 1.44$$

An additional 10% may be added to this figure for operation in the overload region.

Motors are generally not designed with very much extra capacity and therefore are not generally rebuilt for such power changes. A new motor may be required. Check the foundation. Turbines and gears can, like the compressor, be rerated and in some cases have been over designed and will be able to handle the power increase. Whatever your case may be, the driver power capability is a crucial point in the rerate analysis.

PRESSURE

The compressor when originally manufactured was hydrotested at 1.5 times the maximum operating or settle-out pressure. If this pressure is exceeded, the compressor casing should be rehydrotested. Note that on multi-sectional compressors, each section may have been tested at different pressure levels. In such a case each section must be reviewed separately. A rehydrotest should be completed if only one section exceeds the original design pressure.

The ability of the compressor to develop the required pressure rise must be investigated. This is done by calculating the required head using Equation (7.4a). If the gas composition has changed, new values for Z, R, and n must be established (see sample problems in Chapter 7).

$$H_p = Z_1RT_1\left(\frac{n}{n-1}\right)\left[\left(\frac{P_2}{P_1}\right)^{(n-1)/n} - 1\right] \qquad (7.4a)$$

For simplicity, assume projected efficiency to be similar to existing efficiency.

---

*Adapted from "Can You Rerate Your Centrifugal Compressor?", Ron Lapina, Elliott Company, Jeannette, PA, 1975, with permission [20].

Use Equation (2.67) to determine the new speed.

$$N_{\text{rerate}} = N_{\text{original}} \sqrt{H_{\text{rerate}}/H_{\text{original}}} \tag{5.1}$$

This speed may turn out to be too high for impeller stress or rotor dynamics. However, sometimes the manufacturer can increase the head by means of selecting an impeller from a different family designed for high head. Also, these standard impeller designs can sometimes be modified by increasing the diameters. This concept can provide both improved efficiency and increased head as long as the operating speed does not cause excessive impeller stress.

If there is extra space in the compressor for an additional impeller, the head capability can be approximated by

$$H_{p\ \text{rerate}} = H_{p\ \text{original}} [(a + b)/a]$$

where

a = number of impellers in the original rotor
b = number of "blank stages" in the original rotor

Once you have applied this "blank stage factor," go back to the fan law, Equation (2.67) to estimate the new speed required to meet rerate H_p.

SPEED

Speed is a serious consideration. The new speed must not cause high stress in the impellers. The critical speeds of the rotor must be avoided and the rotor must be stable at the maximum operating speed.

It is not really possible for the field engineer to determine the safe operating speed for impellers. Only the manufacturer can determine this. There may be available a new impeller design or improved materials to handle the required speed increase. However, as a starting point you might use 1000 to 1100 fps maximum tip speed for covered wheels and 1200 to 1400 fps for open impellers.

Since the primary factors affecting the critical speeds are bearing span and shaft diameter, the critical speeds are somewhat fixed. However, new bearings or other factors may provide the extra margin to a critical speed if necessary. Again the manufacturer must determine this, along with rotor stability.

Also, remember that the shaft-end stress and torsional criticals must be reviewed. This is usually something left up to the OEM.

As most rerates are field retrofitted, the only testing is done at the site. Thus there is always some question about how the unit will run. An at-speed balance is a good way to gain some added "insurance" that the compressor will operate successfully.

Example 5.2
Consider a "straight through" centrifugal compressor on a dry-air process, with the following data:

Inlet Capacity	= 11,000 icfm
Inlet Temperature	= 90°F
Rated Inlet Pressure	= 14.5 psia
Rated Discharge Pressure	= 55 psia
Rated Power Input	= 1,700 HP
Rated Speed	= 8,100 rpm
Max. Continuous Speed	= 8,500 rpm
First Critical Speed	= 4,800 rpm
Second Critical Speed	= 10,800 rpm
Rated Molecular Weight	= 28.97
$k = c_p/c_v$	= 1.4
Max. Discharge Pressure	= 65 psia

The largest wheel diameter is 22 in, and the cross-sectional drawing of the compressor indicates that the inlet nozzle diameter is 20 in.

The desired rerate is an increase of the inlet capacity to 12,300 icfm and an increase of the discharge pressure to 60 psia, with all other inlet conditions unchanged.

1. Compute the inlet velocity, based on the new inlet volume flow:

$$V_a = 3.06\left(\frac{Q}{D^2}\right) = 3.06\left(\frac{12{,}300}{20^2}\right) = 94 \text{ ft/s}$$

Since this is an acceptable inlet velocity (Figure 6.17), the proposed capacity is feasible.

2. Since the rated inlet conditions have not changed, the increase in weight flow will be proportional to the increase in volume flow, and therefore the power requirement due to the change in volume flow will increase by the same proportion:

$$\frac{\text{GHP}_{\text{rerate}}}{\text{GHP}_{\text{original}}} = \frac{\dot{M}_{\text{rerate}}}{\dot{M}_{\text{original}}} = \frac{Q_{\text{rerate}}}{Q_{\text{original}}}$$
$$= \frac{12{,}300}{11{,}000} = 1.12$$

$$\text{GHP}_{\text{rerate}} = 1.12\ \text{GHP}_{\text{original}}$$
$$= 1.12(1{,}700) = 1{,}900 \text{ HP}$$

Note that, up to this point, the driver will have to be capable of at least

$$(1.1)(1{,}900) = 2,100 \text{ HP}$$

plus 2% excess horsepower if a gear is involved.

3. Since the nameplate maximum-discharge-pressure is 65 psia, the compressor will not have to be hydrostatically retested, provided that the process will not allow the value of 65 psia to be exceeded.
4. The approximate polytropic head can now be calculated for both the original and the rerate conditions from Equation (7.4).

Original:

$$n/(n-1) = k/(k-1)\eta_p = (1.4/1.4-1)(0.76) = 2.66$$

$$H_p = Z_1RT_1(n/(n-1))[(P_2/P_1)^{(n-1)/n} - 1]$$

$$= (1.0)(1{,}545/28.97)(550)(2.66)[(55/14.5)^{1/2.66} - 1]$$

$$= 50{,}700$$

Rerate:

$$H_p = (1.0)(1{,}545/28.97)(550)(2.66)[(60/14.5)^{1/2.66} - 1]$$

$$= 55{,}000$$

The new required speed can be determined from the fan law, Equation (2.67).

$$N_{\text{rerate}} = N_{\text{original}}\sqrt{H_{p\text{ rerate}}/H_{p\text{ original}}} \tag{5.3}$$

$$= 8,100\sqrt{55{,}000/50{,}700} = 8,440 \text{ rpm}$$

The largest wheel diameter is 22 in, therefore, from Equation (6)

$$U - \frac{\pi dN}{720} = \frac{\pi(22)(8{,}440)}{720} = 810 \text{ ft/s}$$

The new rotational speed results in a satisfactory mechanical tip-speed. API states that the second critical speed must be 20% above the highest operating speed. Assuming that the new rerate speed is the highest for the new process, the second critical speed must be at least

$$(1.2)(8{,}440) = 10{,}130 \text{ rpm}$$

The second critical speed (10,800 rpm) is higher than that required and therefore the rotational speed is feasible.

Be sure to check the method used for determining the critical speed. The first critical speed should have been verified in operation if a flexible rotor is used. The second or higher critical(s) may not be accurately known, and the only way to determine this accurately is with a rotor response analysis (a critical speed analysis that considers bearing and support stiffness). While this is being done, rotor stability should also be reviewed.

5. The total increase in gas horsepower can now be determined. The new horsepower will be proportional to the increase in polytropic head and weight flow (in this case, volume flow).

$$\frac{\text{GHP}_{\text{rerate}}}{\text{GHP}_{\text{original}}} = \left(\frac{H_{p\text{ rerate}}}{H_{p\text{ original}}}\right)\left(\frac{Q_{\text{rerate}}}{Q_{\text{original}}}\right) \tag{5.4}$$

$$= \left(\frac{55{,}000}{50{,}700}\right)\left(\frac{12{,}300}{11{,}000}\right) = 1.21$$

$$\text{GHP}_{\text{rerate}} = (1.21)(1{,}700) = 2{,}060 \text{ HP}$$

The driver must therefore be capable of

$$(1.1)(2{,}060) = 2{,}270 \text{ HP}$$

plus 2% if a gear is involved.

Since the inlet velocity, maximum operating pressure and required rotational speed are within satisfactory limits, the rerate is feasible.

SHOP TEST*

The best way to assure equipment quality is to have the compressor performance tested prior to shipment from the factory (Figure 5.5). Although this can be done in the field once the compressor is installed, it is generally difficult to make corrections if necessary, and it is also difficult to demonstrate your concerns to the manufacturer. Timing is usually tight during the installation phase, and once the compressor has been delivered it is more difficult to negotiate with the vendor.

This is true for all compressors, even "off the shelf" varieties. Although the design may be proven, parts can be mismachined or improperly assembled. Custom-built compressors have the added risk of errors in application or design engineering.

ASME Power Test Code (PTC10-1965) has defined three classes of performance tests:

Class 1. This is a test run on the design gas at near design conditions. Generally this applies to air compressors.

Class 2. This covers tests when using the design gas is not practical. Both test and design gas must closely follow perfect gas laws.

Class 3. This test is similar to test 2 in that a different gas is used for test; however, here the gas does not follow the perfect gas law.

The Class 1 test is relatively simple to complete. Be certain of obtaining good data and complete calculations per Chapter 7. Classes 2 and 3 require some special considerations upfront. In determining the proper equivalent conditions the following three items must be reviewed: (1) volume ratio, (2) Mach number, and (3) Reynolds number.

Unfortunately, it is not always possible to match all three parameters with an equivalent test, so some compromise must be made.

On a multi-stage compressor, volume ratio is of utmost concern (see Figure 4.16), and should be closely matched. Per API, the variation between test and actual should not exceed 5%. At other than design volume ratio, downstream stages will not see the design flow and the overall curve will be affected.

To assure the test volume ratio equals the design volume ratio, use Equation (5.5).

$$N_t = N_d \sqrt{\frac{Z_t \mathrm{MW}_d T_t}{Z_d \mathrm{MW}_t T_d}} \sqrt{\frac{[n/(n-1)]_t[(P_2/P_1)^{(n-1)/n} - 1]_t}{[n/(n-1)]_d[(P_2/P_1)^{(n-1)/n} - 1]_d}} \tag{5.5}$$

*Adapted from "Compressor Refresher," Elliott Company, Jeannette, PA, 1975, with permission [11].

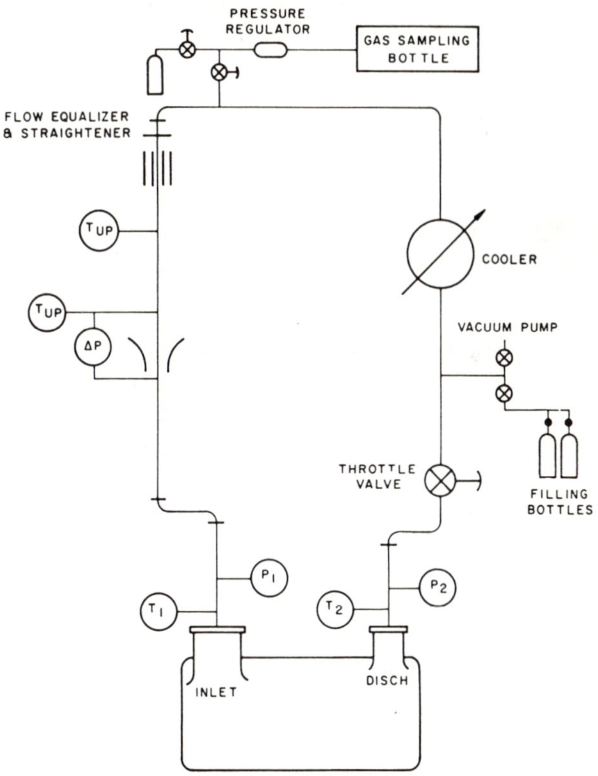

Figure 5.5. Typical shop test [11]. *(Used with permission of Elliott Company, Jeannette, PA.)*

where
d = design conditions
t = test conditions

Note that N_t is proporational to $\mathrm{MW}_d/\mathrm{MW}_t$. Assuming a design MW of 70 and a test MW of 28 (other values fixed), the test speed would be much higher than design. This normally puts the speed beyond the mechanical limits, demonstrating why air or nitrogen cannot be used for equivalent tests for a machine designed for compressing a heavy gas like chlorine or propane.

As demonstrated in Figure 4.9, Mach number determines the capacity of a compressor. Therefore it is critical to assure the test Mach number is within 5% of the design Mach number. The equation for determining test speed to duplicate Mach number is

$$N_t = N_d \sqrt{\frac{(\mathrm{kg}\ ZRT)_t}{(\mathrm{kg}\ ZRT)_d}} \tag{5.6}$$

It is suggested that the test speed slightly exceed the Mach number test speed for conservative results [11]. If possible, select a test gas with k value near the design k value.

With the test speed set with volume ratio and Mach number, we are left with a variation in Reynolds number. As in pipe flow calculations, the Reynolds number affects frictional losses and therefore compressor efficiency. The ASME Power Test Code provides correction factors for variations in Reynolds number.

Except for the variation in actual versus test gas (Classes 2 and 3), the actual shop performance test should closely follow the guidelines presented in Chapter 7. For a Class 1 test, the data is used directly to determine field performance. For Classes 2 and 3, the following is needed to convert test data to actual field conditions.

Capacity

$$Q_d = Q_t(N_d/N_t) \tag{5.7}$$

Head

$$H_d = H_t(\eta_d/\eta_t)(N_d/N_t)^2 \tag{5.8}$$

Efficiency

Note that most users and manufacturers alike agree that the present (1965) PTC-10 guidelines for Reynolds number correction result in too large a correction factor [21]. Therefore Equation (5.9) is offered, which provides half the PTC-10 correction. It is felt that this will provide reasonably accurate results.*

$$\frac{1-\eta_t}{1-\eta_d} = 0.5 + 0.5\left[\left(\frac{\mathrm{Re}_t}{\mathrm{Re}_d}\right)^{0.1}\right]^* \tag{5.9}$$

Machine Reynolds Number

$$\mathrm{Re} = \frac{Ub}{\nu'} \tag{7.17}$$

---

* Note that correction factors are presently being re-evaluated. Look for the latest version of the Power and Test Code for recommended correction factors [21].

6

OPERATION

Although some compressor installations may be relatively simple and knowing the location of the start and stop button is sufficient for operation, it is always best to understand both the operating characteristics (Figure 6.1) of the compressor and how it interacts with the process. Knowing how to properly install, operate, and control a compressor can add years to the useful operating life of the compressor as well as minimize operating costs.

PERFORMANCE CURVES

Before operating a compressor it is important to become familiar with the performance curves.

HEAD/EFFICIENCY

The most versatile of all is the head and efficiency vs. flow curve. This is the basic curve required for describing any compressor (Figure 6.2). Although this curve is useful for small changes in inlet gas conditions, its flexibility is limited. See Figure 4.14.

HORSEPOWER/DISCHARGE PRESSURE

This curve is the most directly usable curve. As long as inlet conditions are matched to the curve, the curve can be read directly without any calculations. The drawback, however, is that inlet conditions are frequently different from the curve, so quite often this curve is useless (Figure 6.3).

PRESSURE RATIO/EFFICIENCY

Pressure ratio and efficiency plotted vs. inlet flow is probably the most useful curve for quickly "ballparking" compressor performance, since inlet pressure is a variable. Note that this curve is only good for the specified suction temperature (Figure 6.4).

If a head curve has been plotted, it is easy to develop a pressure ratio curve using Equation (7.4b).

NOMOGRAPH PLOTS

Nomographs have been a long-time favorite tool of the engineer to save time making tedious calculations. They have been useful even with performance calculations (Figure 6.5). The user should be aware that the accuracy of these nomographs can, however, be very limited. Nomographs can be excellent for rough, quick estimates on the effect of inlet condition changes, but don't expect much more. With all the electronic data reduction equipment available today, it is advisable that the engineer go through the calculations and use the head/efficiency curves described earlier.

Whichever curve is used, it is important to realize that the curves do vary with gas conditions as described earlier. Although the head curve is reasonably accurate for single-stage compressors, significant error can occur for multistage compressors when inlet gas conditions are varied from

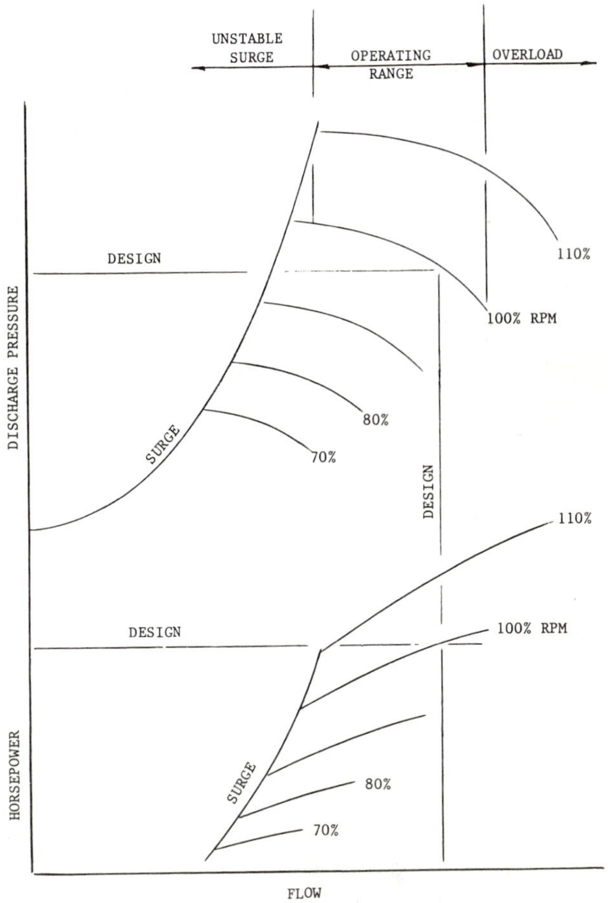

Figure 6.1. Typical performance map.

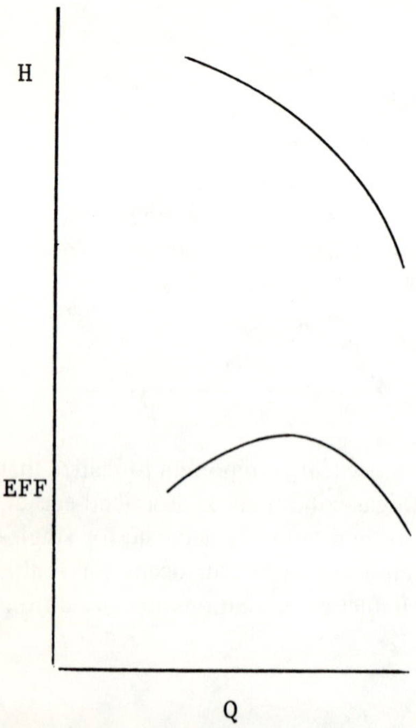

Figure 6.2. Head/efficiency performance curve.

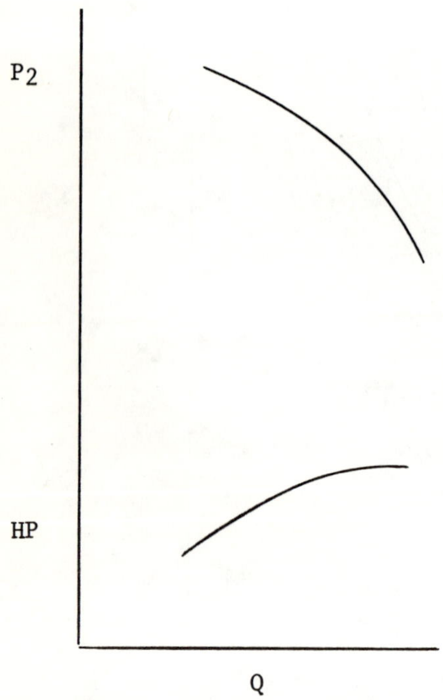

Figure 6.3. Discharge pressure/horsepower performance curve.

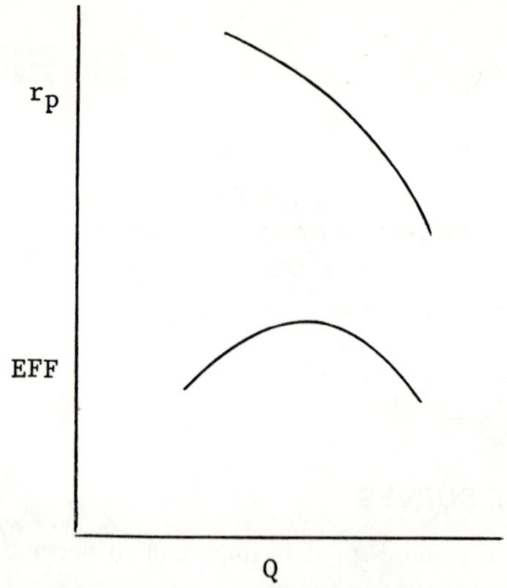

Figure 6.4. Pressure ratio/efficiency curve.

that stated on the performance curve (see "Off-Design Operation," Chapter 4).

When plotting performance data for a variable speed compressor it is wise to make a plot of H/N^2 (or r_p/N^2) vs. Q/N. This will minimize the effects of small (2 to 3%) speed changes (see Figure 6.6).

START-UP*

Normal day-to-day operation of process compressors is relatively straight-forward compared to the transient conditions of start-up and shut-down. During start-up, be it the first time or the 100th, things can change very quickly and it can be difficult to keep on top of the overall operation. This is one of the most crucial times in operating the equipment. The system must be designed, and the operators properly trained, to keep the equipment out of surge, not only during normal operation but also during start-up, shut-down, trip-out situations, or other process abnormalities.

The first start-up may last from several days to many weeks depending on the preparation and complexity of the system. Inlet screens, either temporary or permanent, should be installed on all compressor inlets to ensure that objects are not drawn into the compressor. Caution must be used, however, in the design of inlet screens so they cannot be drawn into the compressor or cause turbulence and affect the compressor performance. Install a ΔP gage across the screen to monitor the condition of the screen.

---

*Adapted from "Compressor Refresher," Elliott Company, Jeannette, PA, 1975, with permission [11].

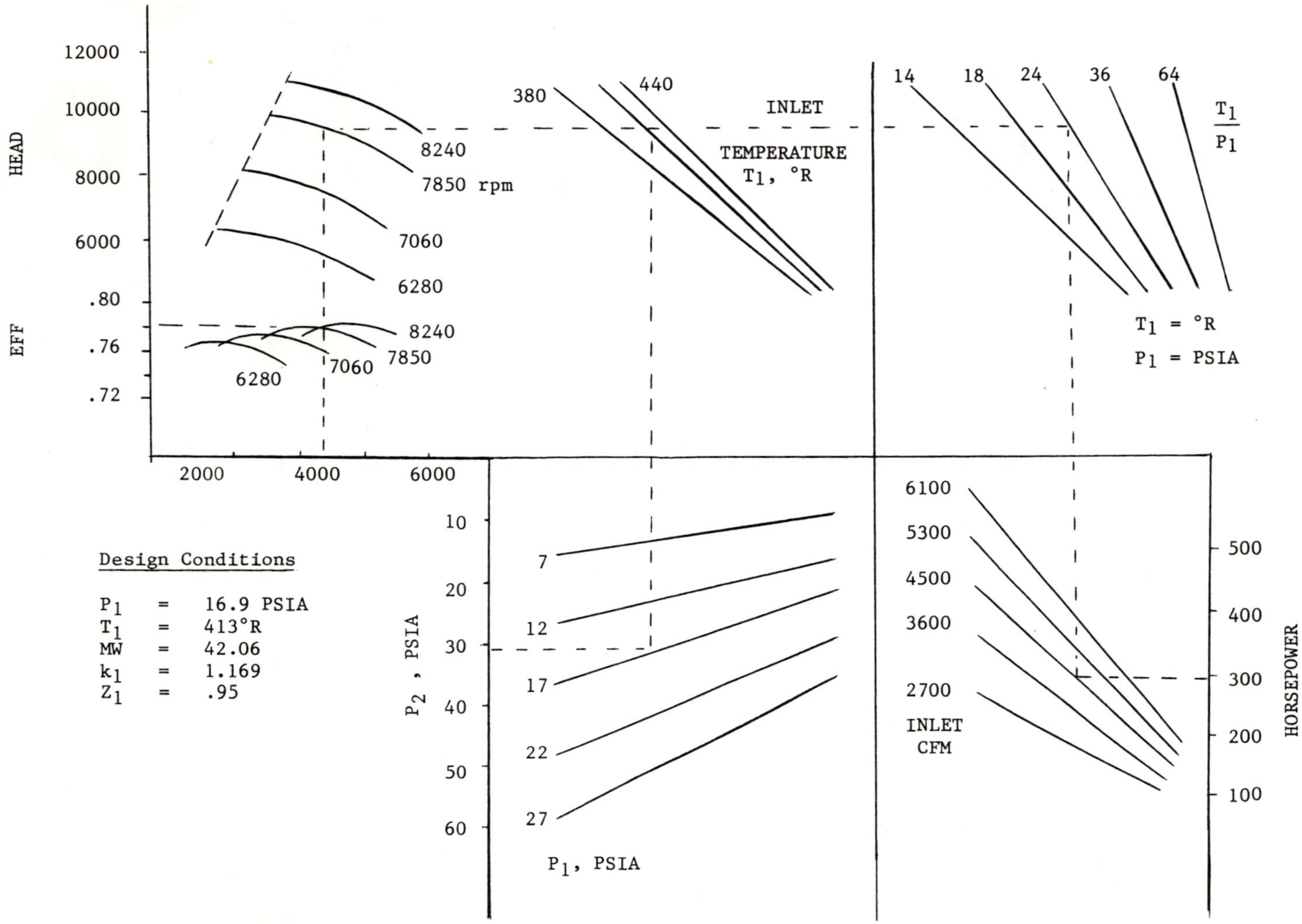

Figure 6.5. Compressor performance in nomograph form.

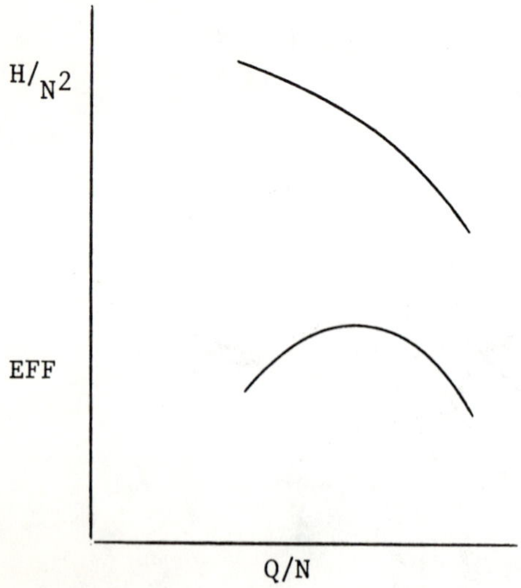

Figure 6.6. Speed compensated curve used for plotting data.

Occasionally, process compressors are operated on air to "dry out" the system before operation on normal process gas. Discharge temperature or seal operation may be the limiting factor here, but occasionally volume ratio effects may prevent operation. For best results, the manufacturer of the equipment should be consulted to review this off-design condition.

When starting a compressor on process, several different methods are available. The method used is a function of the system as well as the driver. For start-up of a steam or gas turbine-driven compressor such as shown in Figure 6.7, coolant would be circulated through the cooler, the bypass or blow-off valve opened, and the unit brought up to operating speed.

The normal procedure for bringing a steam or gas turbine up to speed is usually governed by the manufacturer's recommended starting procedure for the turbine. This should be followed. Typically this involves warmup of the turbine at 500 to 1000 rpm for a few minutes on a gas turbine to one-half hour or more on a steam turbine. Flow control on the compressor is not critical at this time as very little head or pressure is developed. After warming up, the unit's speed is gradually increased. Care must be exercised when approaching criticals, and they should be passed through rapidly. The ability to gradually increase speed is truly a plus feature of the turbine, as it provides the operators with time to make checks and adjustments as the speed is increased. Modern digital controls are a further plus as they can be programmed to not only control speed but also acceleration.

Figure 6.8 shows a typical pressure ratio characteristic which might be experienced as the steam turbine is brought up to speed.

By providing a large bypass or recirculation system around the compressor, the pressure ratio across the unit can be kept low.

Low pressure ratio and corresponding high flow mean very little chance for surge. This area, however, is an operating region where flow separation can occur. On axial compressors, choke flutter can cause serious damage to the blading; therefore, operation of axial compressors far into the overload regions should be avoided.

Once approximately 100% speed is reached (Point "A"), the bypass valve can be slowly closed, causing the

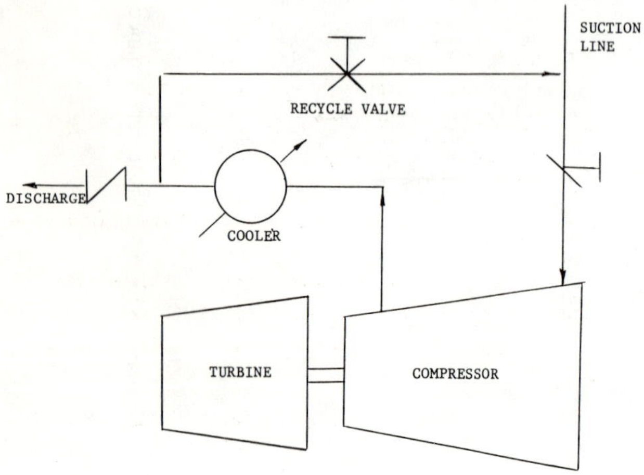

Figure 6.7. Turbine-driven compressor. *(Data from [11].)*

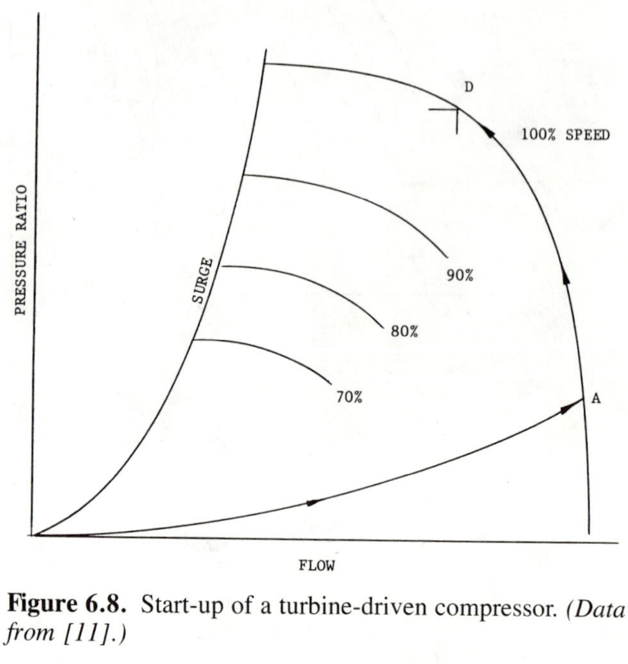

Figure 6.8. Start-up of a turbine-driven compressor. *(Data from [11].)*

compressor performance to ride up the 100% speed line to design point "D".

If the compressor is to follow the path outlined above, the steam turbine must be capable of providing the required additional power. Point "A" (120% torque and 100% speed) corresponds to the peak horsepower typical for closed backward-leaning impellers. While 120% torque is not the worst case possible, it does represent a reasonably safe number for most applications.

Another type of system commonly encountered is the motor-driven centrifugal compressor. Unlike the steam turbine, the motor-driven unit usually has lower starting torque capabilities due to the need for limiting current inrush and heat buildup. Normal start-up time for a motor is limited to less than 20 seconds. Figure 6.9 shows a typical simple motor-driven centrifugal compressor unit.

As in the steam turbine-driven unit, start the coolant circulating in the gas cooler and open the bypass valve prior to starting. In order to minimize current inrush, the inlet butterfly must be closed. After this preparation the machine should come to 100% speed within 10 to 20 seconds.

Preferably the starting characteristic is such that the flow remains to the right of the surge line during the starting cycle (Point B, Figure 6.10). When 100% speed is reached, the butterfly valve must be opened, bringing the compressor away from the surge area and further into the stable operating area.

As in the previous case with the steam turbine, the bypass valve should be wide open before starting the compressor. This puts the compressor discharge at or near suction pressure. In order to allow the compressor to develop

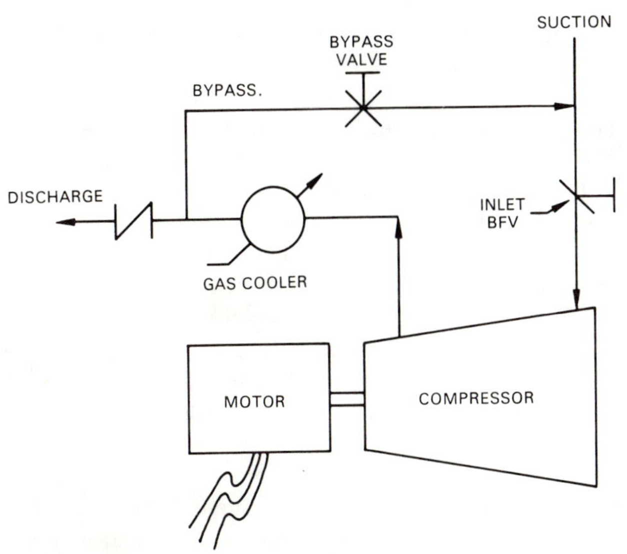

Figure 6.9. Motor-driven compressor [11]. *(Used with permission of Elliott Company, Jeannette, PA.)*

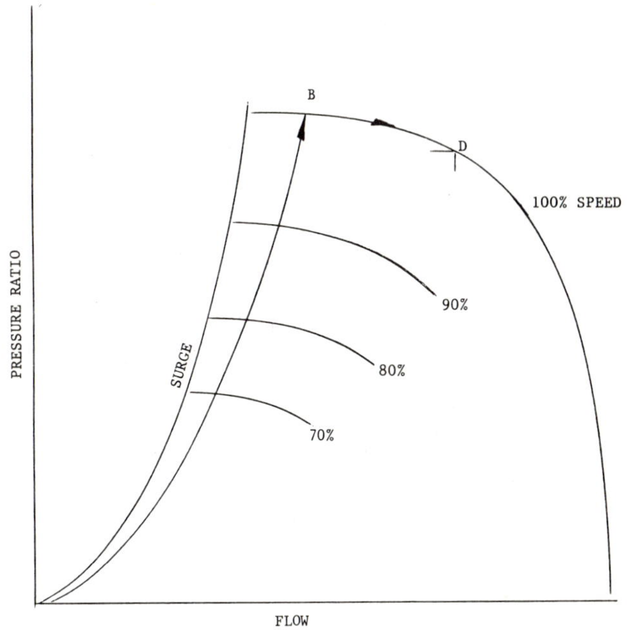

Figure 6.10. Start-up of a motor-driven compressor.

its full design pressure ratio and keep it out of surge, the suction valve must be nearly closed to reduce the suction pressure. The reduced density will reduce the torque required on start-up and permit a smaller motor to be used (Figure 6.11).

Once the compressor is up to speed, the inlet butterfly valve can be opened to increase suction density. The bypass valve can then be closed to raise the discharge pressure, and the suction valve can then be adjusted to maintain proper compressor loading.

On refrigeration units, the discharge flow normally goes to a condenser where the gas is condensed into liquid at a fixed pressure and temperature. In order to get the machine on stream, three things have to happen:

1. Inlet pressure must be kept low to provide low flash temperature.
2. At the same time, suction temperatures must be kept low so that pressure ratios can be realized.
3. Discharge pressure must be reached so condensing starts.

Thus, for a typical refrigeration system, once 30% speed is reached, the speed should be increased rapidly and at the same time the suction drums quenched with liquid in order to provide cooling. At the same time, gas may be blown off to keep suction pressures near design. As speed increases and inlet temperature falls, the blow-off can be closed and the gas diverted to the condenser. Bypasses must be controlled to maintain flows between surge and overload

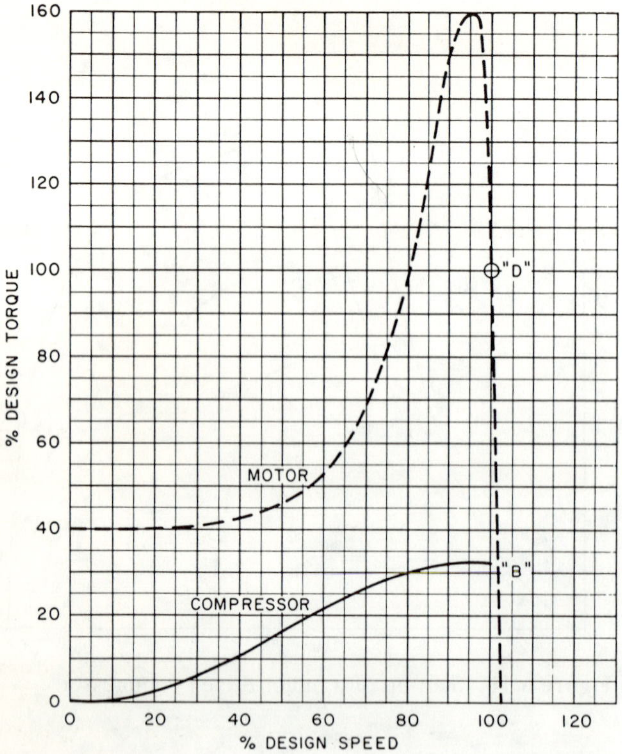

Figure 6.11. Motor/compressor torque curve. Note that the motor torque is higher than the compressor torque. This difference provides the acceleration for start-up. Points B and D correspond to points B and D on Figure 6.10. Torque at point B is low because the suction is throttled to provide low suction density and a corresponding low torque for start-up [11]. *(Used with permission of Elliott Company, Jeannette, PA.)*

so that pressure ratios can be realized as design speed and temperature are approached.

On a compressor with sideloads or one that is intercooled, it is generally a good idea to establish flows and pressures across compressor sections starting at the suction and working toward the discharge.

MECHANICAL FIELD TESTS*

Frequently for a new installation or major rebuild a mechanical test is desired before start-up. Running such a test can prove out the new or rebuilt compressor along with the driver and auxiliaries before plant start-up. The desire is to find bugs in the system in time to minimize delays in plant start-up.

*Adapted from "Guidelines for Mechanical Field Testing of Compressors," J. Dunaway, Elliott Company, Jeannette, PA, 1979, with permission [22].

VACUUM TESTS

Running the compressor with the flanges blanked and 26 to 28 inches of Mercury vacuum pulled on the casing is a good way to prove mechanical integrity of the compressor string. This allows operation of the string to full speed yet with minimal horsepower requirements. Be sure to monitor the temperature at the discharge area, as it should not exceed the manufacturer's maximum temperature limit.

Although the compressor may be in surge during vacuum testing, there should be little temperature buildup and no mechanical damage due to the low energy levels in the gas.

If the unit has oil-lubricated seals, be sure to pull vacuum through the contaminated drains, otherwise the contaminated oil will enter the compressor. Also the seal oil ΔP may require hand control as the overhead tank or regulator may not properly maintain the ΔP. Check the seal bypass orifice or breakdown bushing clearance. These items may require adjustment. Due to the lower pressure, oil ΔP is reduced along with oil flow rates, which may cause overheating of the seals. Check with the manufacturer. Gas face seals may not fully obtain the proper lift at low pressure. Check with the seal supplier to be certain of seal capabilities.

One last thing on vacuum testing. The value of the testing is limited. Aerodynamic cross-coupling effects of internal labyrinth seals will not be demonstrated.

OPEN AIR TESTING

An alternative to the vacuum test is an air test. Normally the compressor is operated with a restricted inlet and an open discharge. As with any off-design operation, it is important to obtain advice from the OEM and stay within the limits specified. Pay particular attention to the discharge temperature and avoid the auto ignition temperature shown in Figure 6.12.

The suction opening can be an orifice plate with a "bird screen" over the opening to keep out debris. Size the orifice with the following procedure. Consider the suction orifice to control suction density, not flow. Assume the flow rate is controlled only by any discharge restriction (piping, check valve, elbows, etc.). With this in mind, and a known discharge pressure of 14.7 psia, assume normal design head and work back to obtain the suction pressure at the first impeller eye, using Equation (7.4c). For example, if a suction pressure of 8 psia were calculated, the suction orifice would then be sized to pass design flow at 6.7 ΔP (14.7–8 psia).

This will assure that the compressor will operate near the design point at design head. Surge and/or high discharge temperature may, however, be unavoidable if the mole weight is significantly different than air (29) due to volume ratio effects (Figure 4.16).

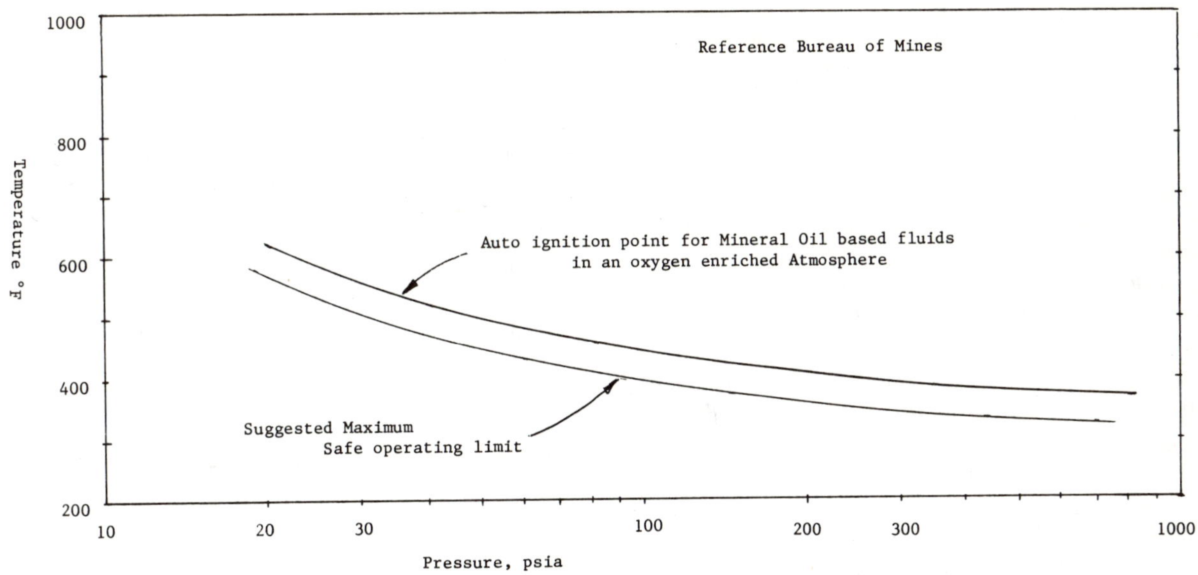

Figure 6.12. Auto-ignition temperature. *(Data from [22].)*

Note that as in the vacuum test, there may be some problems with the seals. Assure proper oil flow and watch for overheating of the seals. Buffer seals to minimize oil entering the machine.

Depending on the design density of the process gas, the air can result in power requirements above normal. If so, full speed may not be reached. Be sure to closely monitor the driver.

Verification of aeroinstability effects requires duplication of gas density. This may not occur for the air test described above.

One special note of caution on an air test. Too often someone may desire to run an air test and simply remove the inlet and discharge piping and start up the unit on air.

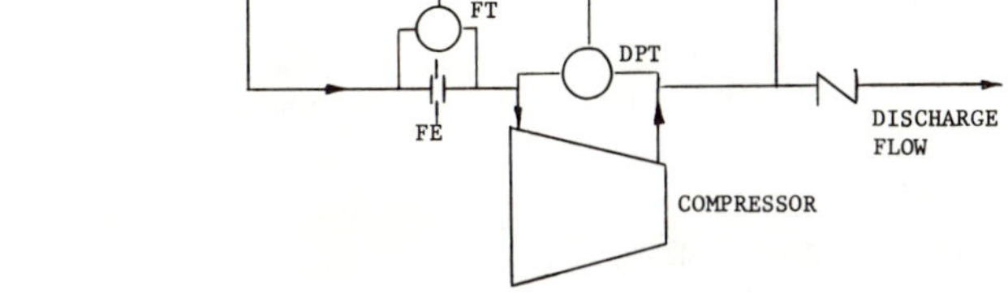

Figure 6.13. Typical anti-surge system [18]. *(Used with permission of Elliott Company, Jeannette, PA.)*

This is a big mistake! The compressor is attempting to make design head. Also, the pressure at the discharge is 0 (14.7 psia). This means that the first wheels are producing head, while the last wheel is turbinizing. Serious damage has occurred under this condition and it should be avoided.

Remember to allow the compressor to develop its design head. Any restriction at the discharge will limit flow, so this must be minimized. Reduce the suction density by throttling the inlet.

FULL LOAD TEST

Sometimes it may be possible to run a loop test in the field with normal process gas. If properly designed, the compressor will have a recycle loop capable of passing design flow. This loop will include a cooler, control valve, and orifice for measuring flow (Figure 6.13). This situation will provide the best mechanical/aero test possible. All systems should be operating near normal conditions, thus closely simulating operation during plant operating conditions. Also, since process gas is being used, aerodynamic performance can also be verified.

AVOIDING SURGE*

The most important item for protecting the compressor from surge operation is the anti-surge control system. This control system should maintain a minimum volume of flow through the machine so that the surge condition is

* Adapted from "Compressor Performance," R. Salisbury, Elliott Company, Jeannette, PA, 1986, with permission [18].

never encountered. This is achieved by bleeding flow from the discharge of the machine to maintain a minimum inlet flow. This flow can either be dumped to atmosphere or recirculated back into the inlet of the compressor. In the latter case, it must be cooled to the normal inlet temperature. For most applications a simple control based on a flow differential is adequate for this function. However, on machines where the speed or the gas conditions are variable, the control may have to be more sophisticated to insure proper operation under all conditions. This is frequently achieved by modulating the control with a signal for pressure, temperature, speed, or a combination of parameters.

Provisions must be made for start-up and trip-out of the machine with sufficient through flow to prevent surging and excessive heating of the inlet gas.

On any control scheme a trip-out of the driver should be interlocked to open the anti-surge valve within 3 seconds and allow the machine to coast to a stop with this line open. Otherwise, the machine could be surging constantly while dropping down in speed. This is particularly important for axial compressors and also for high-pressure, high-horsepower centrifugal applications. A typical anti-surge control system is shown in Figure 6.13.

A description of the function of each component is as follows:

- *FE* The flow element is usually an orifice located in the compressor suction, although it can be a venturi or calibrated inlet such as those used in axial compressors. Its purpose is to cause a temporary pressure drop in the flowing medium in order to determine the flow rate by measuring the difference of static pressures before and after the flow-measuring element.
- *FT* The flow transmitter is a differential pressure transmitter which measures the pressure drop across the flow element and transmits a signal that is proportional to flow squared.
- *DPT* The differential pressure transmitter measures the differential pressure across the compressor and transmits an output signal that is proportional to the measured pressure differential.
- *FX* The ratio station receives the signal from the flow transmitter and multiplies the signal by a constant. This constant is the slope of the control line.
- *FZ* The bias station receives the signal from *FX*, the ratio station, and biases the surge control line.

The ratio station must have both ratio and bias adjustment to enable the control line to be placed as parallel to the compressor surge line as possible. (See Figure 6.14.)

$$\Delta P = Ch + b \tag{6.1}$$

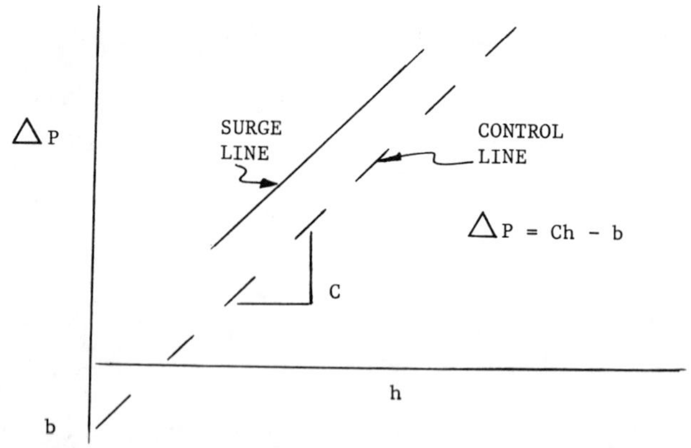

Figure 6.14. Surge control line [18]. *(Used with permission of Elliott Company, Jeannette, PA.)*

where

ΔP = Calculated compressor differential signal
C = Control line slope (ratio signal)
h = Inlet orifice differential signal measured by FT
b = Control line bias (could be zero)

- *FIC* The surge controller is a flow control device which compares the calculated output of FZ to the measured ΔP output of the DPT with ΔP as defined above.

  When the calculated ΔP is greater than the measured ΔP, the compressor is operating to the right of the control line. When the calculated ΔP is equal to or less than the measured ΔP, the compressor is on or to the left of the control line, and the surge controller functions as a flow controller and opens the anti-surge valve as necessary to maintain operation of the compressor on this surge control line.

  For rapid flow changes, the response of the control system must be rapid to prevent surge.
- *LAG* This device functions to enable the surge controller to open the recycle valve quickly, while providing a slow closure rate. This feature provides stability between the anti-surge control system and the process by minimizing the hunting effect between control system and recycle valve.
- *LX* The low signal selector is set up for two inputs and one output. The inputs are a 100% signal valve and the surge controller output signal. The output of the low selector is sent to the recycle valve as well as back to the surge controller in the form of a feedback signal. This prevents the surge controller from winding up. Windup of the controller penalizes the reaction time of the anti-surge control system.
- *FCV* The anti-surge recycle valve functions to prevent surge by recycling flow from the compressor discharge back to the compressor inlet.

LIQUIDS*

One of the most potentially damaging occurrences for a compressor is the ingestion of liquid with the process gas stream. Liquids condensing in the recycle line can minimize the effectiveness of any anti-surge system by creating a blockage in the line. The full range of operation should be studied to avoid having liquids enter the compressor during normal operation and upsets.

1. Trim cooling water or other process conditions to keep the compressor inlet conditions above the liquefaction points for any gas constituent.
2. Heat trace, bleed off, or purge normally stagnant lines when liquids form as the stagnant gas cools down to ambient temperatures.
3. Recycle lines should re-enter main gas stream either upstream of, or at the inlet knockout drum.
4. If any potential for liquid formation exists upstream or downstream of the compressor, drains and level indicators should be provided at all low spots of piping and vessels. This will allow routine checking for liquids and draining as required.
5. After shutdown be sure all liquids formed by cooldown of the stagnant process gas are drained away before the compressor is restarted.

*Adapted from "Compressor Refresher," Elliott Company, Jeannette, PA, 1975, with permission [11].

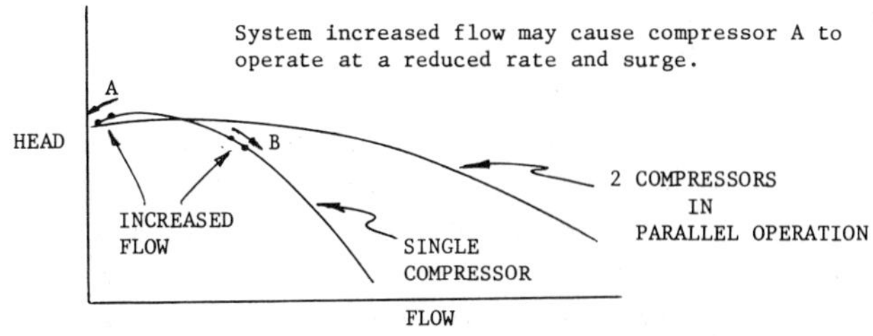

Figure 6.15. Parallel operation. *(Data from [9, p. 277].)*

PARALLEL OPERATION

A common method of increasing capacity of a system while enhancing reliability is using two or more compressors operating in parallel. A combined characteristic of two identical units in parallel is shown in Figure 6.15. Since the "identical" units are always somewhat different and system resistance varies, it is feasible that both units will not be operating at the same point on the curve. It is possible for one unit to be in surge, while the other is not. It is therefore recommended that each unit have a separate anti-surge system.

DOUBLE FLOW COMPRESSORS

In double flow compressors (Figure 6.16) a factor that must be considered is the design of the inlet piping to achieve a well balanced, distortion-free flow into each inlet of the compressor. Otherwise, as with the parallel operation described above, the compressor will not handle equal flow rates on each side and premature surging may occur before the anti-surge control is activated.

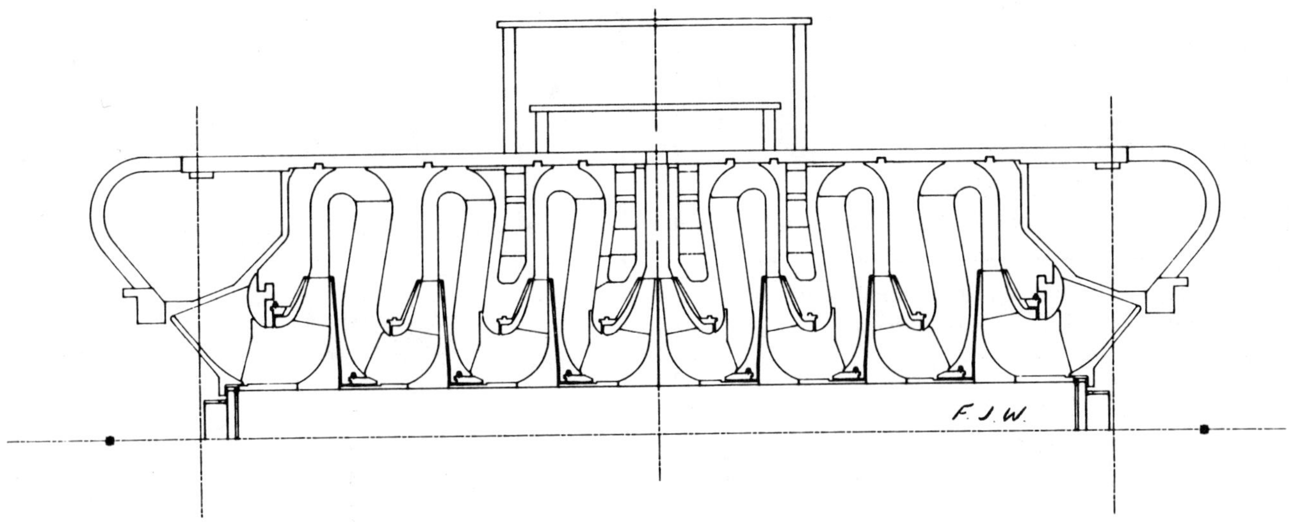

Figure 6.16. Double-flow compressor.

INLET PIPING*

Compressor performance is very dependent on obtaining a uniform flow distribution to the impellers. Great pains are taken by the compressor designer to assure proper flow distribution to intermediate impellers. Although inlet guide vanes may exist on a compressor, this alone does not assure proper flow distribution to the first stage impeller. Compressors are designed assuming a relatively uniform velocity profile at the compressor inlet flange.

Although ASME goes to some detail in describing upstream straight-run requirements for orifice meters, the requirement for compressors is very simply stated. The straight-run requirement for axial inlet compressors is 10 pipe diameters and for non-axial inlets, 3 diameters. Additionally, if the velocity pressure exceeds 1% of the static pressure, a flow equalizer must be used at the exit of the elbow upstream of the compressor inlet flange (Figure 6.17).

$$\frac{P_v}{P_1} \leq .01 \tag{6.2}$$

where

$$P_v = \frac{P_1 V_1^2}{2gZ\,RT_1} \tag{6.3}$$

Values for $P_v/P_1 = .01$ are shown in Figure 6.17. Note that for 100°F air (MW = 29) the maximum inlet velocity for a three-diameter straight run without an equalizer is 140 fps. For propane (MW = 44) at 0°F, the maximum velocity would be 100 fps.

According to Hackel and King [23] the ASME guideline is adequate but could be modified to call for a reduced straight run for smaller P_v/P_1 values, and greater straight run for larger P_v/P_1 values. Also, additional length of straight run should be required for compound piping arrangements.

For a base case of one elbow in a plane parallel to the compressor axis and for a radial inlet with 50°F gas, Figure 6.18 gives the minimum straight run required.

To correct for other suction temperatures, use the following equation to find the equivalent velocity for 50°F. First calculate the actual velocity V_1.

$$V_{50F} = \frac{22.6V_1}{\sqrt{T_1}} \quad \text{where } T_1 = \ ^\circ R \tag{6.4}$$

For axial inlets and/or other inlet piping arrangements use Figure 6.18 along with the multipliers provided in Table 6.1.

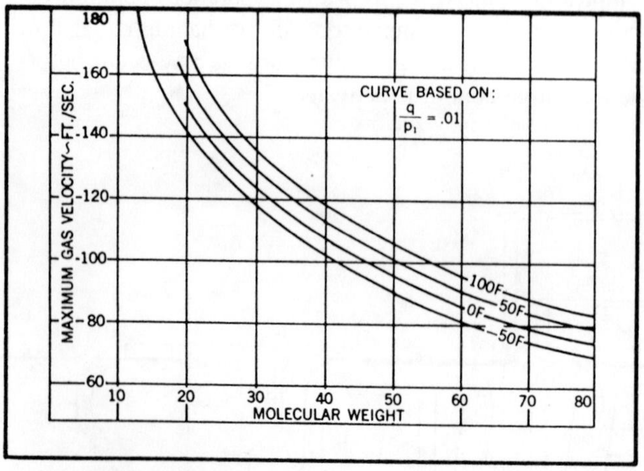

Figure 6.17. Maximum velocity allowed at suction flange based on Equation 6.2 [23]. *(Used with permission of Elliott Company, Jeannette, PA.)*

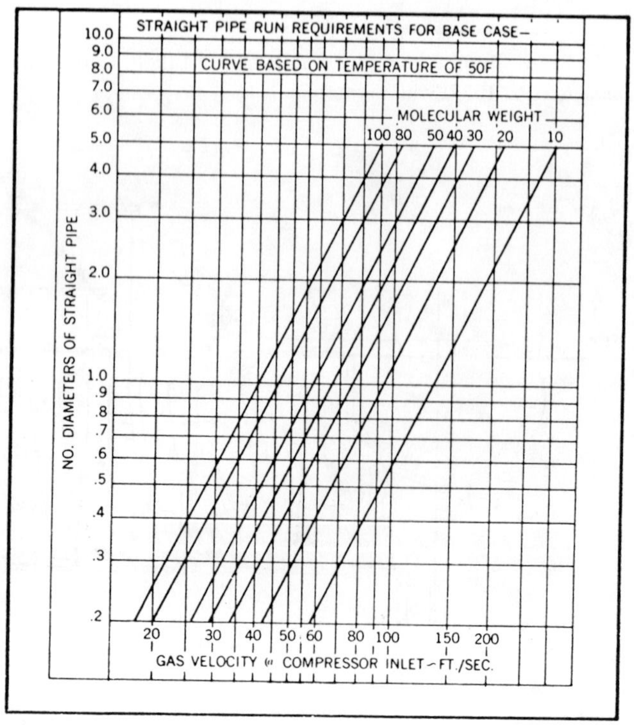

Figure 6.18. Straight pipe run requirements for Case 1 in Table 6.1 [23]. *(Used with permission of Elliott Company, Jeannette, PA.)*

*Adapted from "Centrifugal Compressor Inlet Piping—A Practical Guide," Ross Hackel and Raymond King, Elliott Company, Jeannette, PA, 1977, with permission [23].

TABLE 6.1 Multipliers for Various Inlet Piping Arrangements*

Piping Configuration	Multiplier Radial Inlet
One long radius elbow in a plane parallel to compressor shaft	1.0
One long radius elbow in a plane 90° to compressor shaft	1.5
Two elbows at 90° to each other	
Elbow nearest compressor in a plane parallel to shaft	1.75
Above, only elbow normal to shaft	2.0
Butterfly valve before an elbow	
Valve axis normal to compressor shaft	1.5 to 3.0
Valve axis parallel to compressor shaft	2.0 to 4.0
Butterfly valve in straight run entering compressor inlet	
Valve axis normal to rotor	1.5 to 3.0
Valve axis parallel to rotor	2.0 to 4.0
Two elbows in the same plane parallel to compressor shaft	1.15
Two elbows in same plane 90° to rotor	1.75
Gate valve wide open	1.0
Swing check valve balanced	1.25
Axial inlet	1.25

*Use with Equation (6.4) and Figure 6.18 to determine required straight run of piping for a given arrangement. (From [23].)

EXAMPLE 6.1

Consider a gas with a MW of 25, inlet temperature of 85°F and suction velocity of 100 fps. The piping configuration is a radial inlet multi-stage compressor with two elbows in different planes upstream of the compressor. The elbow nearest the compressor is in a plane parallel to the rotor. Upstream of the two elbows is a butterfly valve.

Figure 6.18 gives a straight-run requirement of 1.45 diameters. The multiplier for the two elbows is 1.75 and butterfly value factor is 1.5.

Multiplying

$$1.45 \times 1.75 \times 1.5 = 3.8$$

Four pipe diameters of straight run are required for this application.

7

FIELD PERFORMANCE TESTING

This chapter provides a guideline for field performance testing of compressors. Included are procedures, equations, sample calculations, and data trending procedures.

Before attempting a performance test, review the following check list and be certain all the required data can be obtained. As a reference, see your instruction book for design conditions. (See Appendix A, Table A.4.)

- *Pressure and temperature at each flange*
- *Mass flow rate*
- *Gas properties*
- *Equipment speed*
- *Driver power*
- *Compressor and driver mechanical losses*

Field performance test procedures should be in accordance with ASME PTC10–1965 Compressors and Exhausters within practical limits. The method outlined here is for routine performance evaluation only. This procedure is not necessarily sufficient for an OEM acceptance test. If an acceptance test is to be performed, details should first be worked out with the manufacturer.

Compressor shaft horsepower is best determined by adding the enthalpy rise gas horsepower to standard values of bearing and seal mechanical power losses.

Mass flow is determined by using the process flowmeter. The flow rate is adjusted to account for variations from meter design flow conditions. Mass flow is checked by direct calculation from flowmeter upstream conditions and differential pressure using applicable ASME flow code equations.

Test point readings should not be taken until such time as the compressor is shown to be in equilibrium. Equilibrium is defined as the condition in which the discharge temperature does not vary more than 1°F over a five-minute period at constant inlet temperature.

Upon achieving equilibrium, three complete scans of data readings per data point is taken over a twenty- to thirty-minute period and averaged for calculations.

It is recommended that a minimum of three and preferably five data points be taken to establish the performance curve shape. Take one point at the design or normal operating point, at 95% of this value, 105%, 110%, and 90%.

A gas sample should be taken at the beginning and end of the test point. Gas sampling at the inlet can minimize condensation in the sample bottle. However, it is wise to sample at both suction and discharge. The recommended method of gas analysis is by gas chromatograph.

An estimate of overall test accuracy should be made by performing a power balance. Measured driver-delivered power minus gear (if applicable) power losses can be compared to calculated compressor-absorbed power. The overall accuracy of the test is no better than the accuracy indicated by this difference.

GAS SAMPLING

A stainless steel sampling cylinder should be used. It should be at least 300 ml in size and have straight cylinder valves on both ends. The pressure rating of the cylinder should be high enough so that it can withstand full system pressure. To prevent gas condensation (the walls of the cylinder and tubing are relatively cool) during the sampling process, the cylinder and lines leading from the process pipe should be insulated and heat traced. The sampling container should be purged for about 5 minutes before closing the valves and trapping the gas at process pressure.

The containers are then transported to the laboratory. During the transportation the samples cool and therefore drop in pressure. The cooling of the sample may result in condensation of some of the gas. This condensation must be gasified before feeding into the gas chromatograph. The only sure way to do this is to heat the sample to or above the sample temperature. It may be wise to then bleed off some pressure before feeding the gas chromatograph. Plot this out on a Mollier diagram of the gas. This brings you further away from the dew point and provides further assurance of avoiding condensation.

A cross-check on your accuracy can be made by checking the weight of a separate sample. This sample container should be at a vacuum and heated before filling with the process gas. Weigh the sample container evacuated and with the sample. Knowing the weight and volume will give you the specific volume. Check this against the calculated specific volume for temperature and pressure of the sample point using the composition given by the gas chromatograph.

Of course this procedure is not necessary in all cases, such as low mole weight mixtures. For heavy hydrocarbons, however, it is essential that the above procedure be followed. Errors in gas analysis can give significant errors in performance.

INSTRUMENTATION*

General instrumentation requirements are shown in Figure 7.1. All temperature and pressure indicators must be dual, i.e., a minimum of two independent instruments per location.

Ideally, such as under development testing conditions, the static pressure tap hole should be very small (approximately 1/5) compared to the boundary layer thickness [25]. But on a more practical note, the static pressure connection should have a pressure tap hole .25 inch in diameter but no greater than 0.5 inch, deburred and smooth on its inside edge, with a sharp corner. A smaller hole will collect dirt or condensate. Note that the hole must be deburred but have

*Adapted from "Field Performance Testing," D. Bensema, Elliott Company, Jeannette, PA, 1986, with permission [24].

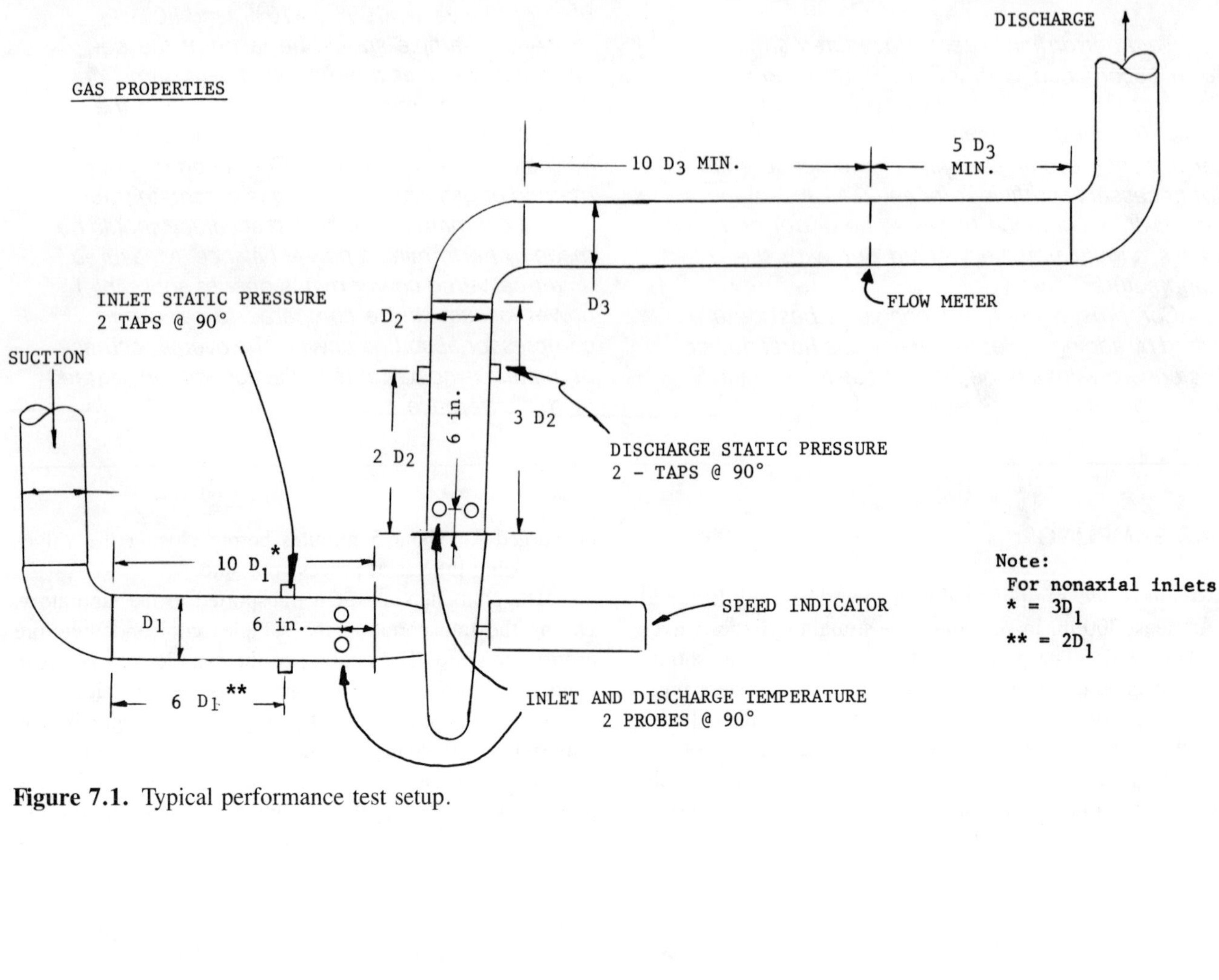

Figure 7.1. Typical performance test setup.

a sharp corner on the measurement side. Take care that the deburring procedure does not round the edges, as this will give erroneous readings (Figure 7.2).

If possible, all pressures below 20 psig should be measured with a vertical-type fluid manometer of single- or double-leg design. Manometer fluids must be chemically stable when in contact with the test gases, and specific gravity must be measured before and after the test. If safety regulations do not permit the use of manometers, test gauges should be used.

All pressures above 20 psig at compressor flanges and flowmeter devices should be measured using quality gages, 6-inch or larger diameter, having a 0.25% sensitivity, and a maximum error of 0.5% full scale.

Pressure readings during testing should be at mid-scale or greater. Mounting should be on a vibration-free local panel, connected with pressure lines of at least 1/2 inch I.D. tubing; lines will continually slope down toward the unit to automatically drain any condensate. Block vent valves should be mounted at the gages to facilitate their in-place calibration. Calibration using a certified dead weight tester is preferred. Suggested arrangements are shown in Figures 7.3 and 7.4.

Temperatures are to be measured using a thermocouple or RTD system having a sensitivity and readability of 0.5°F

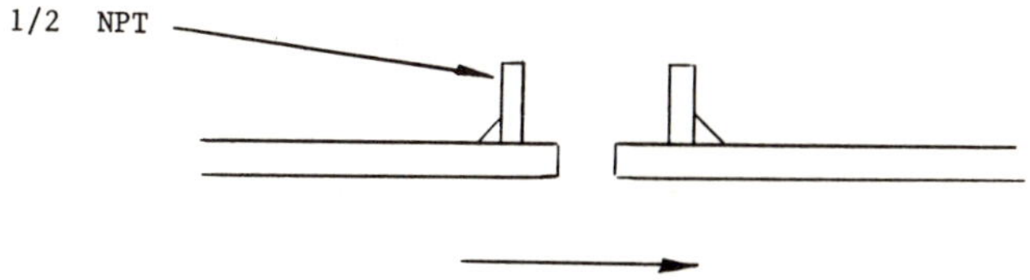

Figure 7.2. Static pressure tap. The hole should be $\frac{1}{4}$ to $\frac{1}{2}$ inch in diameter. It should be deburred but have a sharp edge.

and an accuracy within 1°F. Care should be taken to avoid intermediate T–C junctions at terminals and switch boxes with a thermocouple system. Glass-stem thermometers are generally unacceptable for safety reasons. The temperature-sensing portion of the probe must be immersed into the flow to a depth of 1/3 to 1/2 the pipe diameter. The temperature-sensing elements should be in intimate thermal contact if using wells, utilizing a suitable heat transfer filling media, such as graphite paste. Stem conduction errors can be further minimized by wrapping the stem and well with fiberglass or wool insulation.

Speed should be determined utilizing two independent systems, one being the compressor key-phasor with calibrated digital readout with 0.25% or better system accuracy.

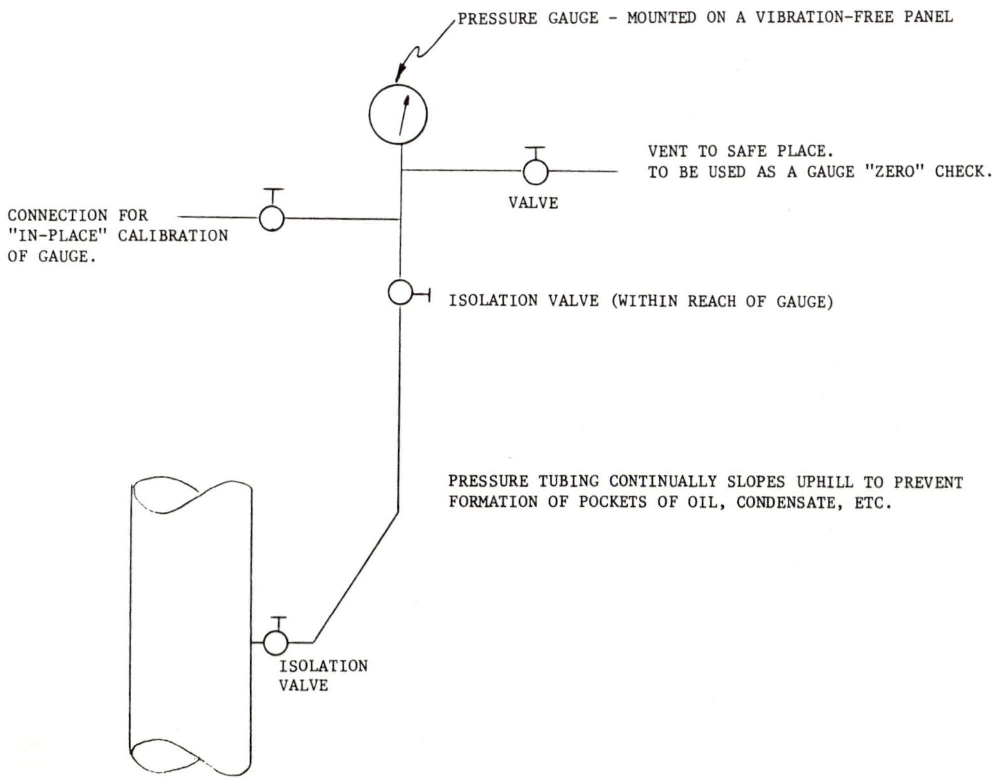

Figure 7.3. Typical instrument line with gauges mounted above pressure tap. *(Data from [24].)*

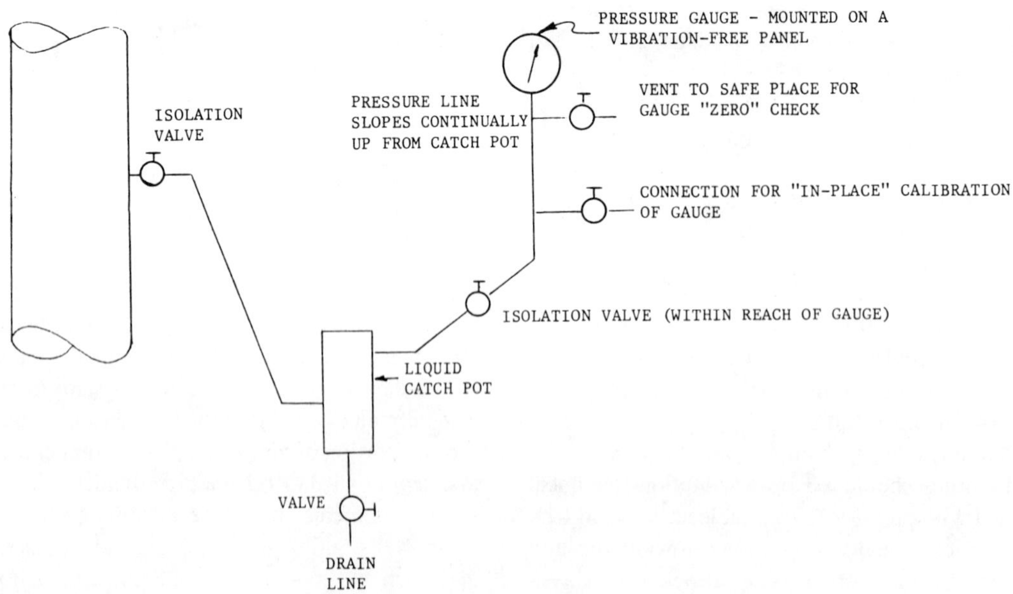

Figure 7.4. Typical instrument line installation where pressure line cannot be sloped continuously upward from pressure tap to gauge. *(Data from [24].)*

Pressure taps should be spaced 90° apart. On horizontal runs of pipe, pressure taps must be in the upper half of the pipe only.

Flow rates derived from the process flow indicator should be checked by direct computation of mass flow rates through the metering device. For this reason, metering device upstream temperature, upstream pressure, and differential pressure must also be recorded.

SIDELOADS AND EXTRACTIONS

Preferred instrument locations are as shown in Figure 7.5. Sideload and extraction lines, if applicable, are to be treated as inlet and discharge lines, respectively. If existing instrument tapping points must be used, care must be taken that those used are reasonably close to the compressor flanges. There must be no valves, strainers, silencers, or other sources of significant pressure drop between the pressure tap points and the compressor flanges.

Evaluation of sideload and extraction compressor performance is difficult and subject to significant inaccuracies unless internal pressure and temperature probes at the sideload and/or extraction lines are available.

Pressure and temperature taps can easily be added to a horizontally split compressor as shown in Figure 7.5 by drilling and tapping the casing in the return channel crossover area. A minimum of two each pressure and temperature taps should be used. For barrel-type compressors, use the casing drain in this area if available. Although neither of these methods will provide precise data, they will obtain good relative data with which you can track the relative performance of your compressor. This data can become more realistic if a total mass balance and power balance is done on the system to determine accuracy of the test.

If precise data are required, such as for an acceptance test, it may be necessary to run temperature and pressure probes to the end of the return channel as shown in Figure 7.5.

Once pressures and temperatures are known at the discharge of Section I, a mixing calculation is required to establish suction conditions for the next section (see Figure 7.5).

$$P_1 = P_2 = P_3 \tag{7.1}$$

where

1 = Discharge of Section I
2 = Sideload Condition
3 = Mixed Suction to Section II

$$\dot{M}_1 h_1 + \dot{M}_2 h_2 = \dot{M}_3 h_3 \tag{7.2}$$

where $\dot{M}_1 + \dot{M}_2 = \dot{M}_3$.

T_3 is then found by working back through the Gas Properties or Mollier Diagram knowing h_3 and P_3.

T_3 may be very accurately approximated by

$$T_3 = \frac{\dot{M}_1 T_1 + \dot{M}_2 T_2}{\dot{M}_1 + \dot{M}_2} \tag{7.3}$$

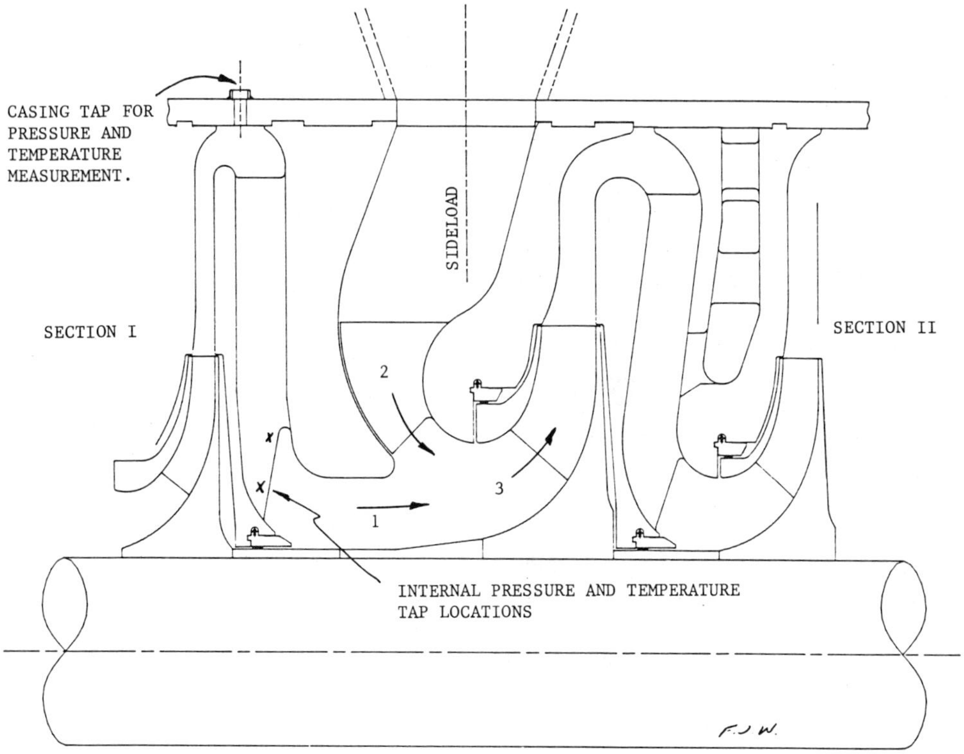

Figure 7.5. Pressure and temperature taps for a sideload nozzle.

SPECIAL DATA REDUCTION FOR SIDELOADS

Special data reduction techniques can be used on sideload and extraction compressors where internal pressures and temperatures are not available. Internal pressures can be estimated from flange pressures, gas velocity through the compressor nozzle, and standard pressure drop loss coefficients for a given sideload or extraction nozzle design.

Internal gas temperature at the discharge of each section is also required to determine sectional performance. This can be accomplished through an iterative process which makes use of predicted work curves for each section. The procedure begins for a given test point by establishing the inlet volume flow for Section I. From the predicted work curves, the estimated work input is obtained. These data, along with the internal pressure determined above, are used to establish the estimated discharge temperature for Section I.

A Mollier diagram or data with the enthalpy as a function of pressure and temperature is used. The sideload flow, as measured on site, will then be mixed with the calculated discharge flow from Section I to establish the inlet flow to Section II. This procedure is then repeated for each following compressor section using its respective work input plot. The test on the validity of the work input curves is made by comparing the calculated final discharge temperature with the measured final discharge temperature. If these two temperatures agree, the assumption is made that the correct work input has been used. If, however, the two temperatures do not duplicate one another within two degrees ($\pm 2°$), the work input curves for each section are varied by the same percentage, and the process is repeated.

Once the sectional inlet and discharge conditions are determined, the sectional heads and efficiences can be calculated.

INSTRUMENT CALIBRATION*

In general, all instruments used for the measurement of temperature, pressure, flow, and speed should be calibrated by comparison with appropriate standards before the test. General recommendations for calibration procedures are outlined below. A comprehensive log book should be maintained for all calibrations. Pressure gage calibration should state actual deadweight pressure and indicated gage value.

Pressure gages should be check calibrated against deadweight standards throughout the range. Calibration using both increasing and decreasing pressure signals should

*Adapted from "Field Performance Testing," D. Bensema, Elliott Company, Jeannette, PA, 1986, with permission [24].

be done to check for hysteresis. Gages not within 0.5% error of full scale should not be used. Needles should not be changed or adjusted. The gages must have a readable sensitivity to 0.25%.

There should also be a check on the accuracy of the thermocouple system (lead wires, reference junctions, readout devices) for each thermocouple. One method of accomplishing this is to read voltage output of the thermocouple locally, and then compare this to the remote reading of thermocouple output.

It is also recommended that the accuracy of the thermocouple itself be checked by subjecting it to varying temperatures and comparing its output to a reference standard. The thermocouple should be checked throughout its operating temperature range. The thermocouple system should have a readable sensitivity to 0.5°F and an accuracy within 1°F.

Calibration of the flowmeter differential pressure transmitter can be verified by impressing a known differential pressure across it and measuring its output. Finally, an overall system check can be made by impressing a differential pressure across the transmitter and reading the final control room output. The flowmetering device should be removed, and its dimensions should be checked, recorded, and compared to design criteria.

The tolerance for the measurement of compressor speed should not exceed 0.25%. Use of two independent instruments, one to provide a check on the other, is recommended. Electronic tachometers must be checked.

CALCULATION PROCEDURES*

GENERAL

Data reduction involves the equations shown below. Performance parameters are calculated using actual gas properties based upon the results of the on-site gas analysis. Performance parameters are calculated for each section of compression based on flange-to-flange data. The gas properties procedure or Mollier method is the preferred method for determining polytropic head and efficiency.

HEAD

1. Polytropic

$$H_p = Z_1 R T_1 \left(\frac{n}{n-1}\right)\left[\left(\frac{P_2}{P_1}\right)^{(n-1)/n} - 1\right] \tag{7.4a}$$

where

$$n = \frac{\ln(P_2/P_1)}{\ln(v_1/v_2)}$$

$$\frac{v_1}{v_2} = \frac{Z_1 R T_1/144 P_1}{Z_2 R T_2/144 P_2} = \frac{Z_1 T_1 P_2}{Z_2 T_2 P_1}$$

$$n = \frac{\ln(P_2/P_1)}{\ln(Z_1 T_1 P_2/Z_2 T_2 P_1)}$$

If head is known, and r_p or P_2 is desired,

$$\frac{P_2}{P_1} = \left[\frac{H_p}{Z_1 R T_1\,[n/(n-1)]} + 1\right]^{n/(n-1)} \tag{7.4b}$$

$$P_2 = \left[\frac{H_p}{Z_1 R T_1\,[n/(n-1)]} + 1\right]^{n/(n-1)} (P_1) \tag{7.4c}$$

See also Equations (2.36) and (2.37).

2. Adiabatic (Use for low-pressure air compressors.)

$$H_{ad} = R T_1 \left(\frac{k}{k-1}\right)\left[\left(\frac{P_2}{P_1}\right)^{(k-1)/k} - 1\right] \tag{7.5a}$$

where

H_{ad} = Adiabatic Head, Feet

$k = c_p/c_v$

$$\frac{P_2}{P_1} = \left[\frac{H_{ad}}{R T_1\,[k/(k-1)]} + 1\right]^{k/(k-1)} \tag{7.5b}$$

$$P_2 = \left[\frac{H_{ad}}{R T_1\,[k/(k-1)]} + 1\right]^{k/(k-1)} (P_1) \tag{7.5c}$$

See also Equation (2.38).

EFFICIENCY

1. Polytropic

$$\eta_p = \frac{H_p}{(h_2 - h_1)\,778.16} \tag{7.6}$$

where

H_p = Polytropic head, in feet
h_2 = Enthalpy at discharge conditions in BTU/LB
h_1 = Enthalpy at inlet conditions in BTU/LB

2. Polytropic—Constant k

$$\eta_p = \frac{(k-1)/k}{(n-1)/n} \tag{7.7}$$

*Adapted from "Compressor Performance," R. Salisbury, Elliott Company, Jeannette, PA, 1986, with permission [18], and "Field Performance Testing," D. Bensema, Elliott Company, Jeannette, PA, 1986, with permission [24].

3. Adiabatic (Air Compressors)

$$\eta_{ad} = \frac{T_1[(P_2/P_1)^{(k-1)/k} - 1]}{T_2 - T_1} \quad (7.8)$$

FLOW

Basic flow measurement equations, ASME PTC19.5-1971, "Fluid Meters." See also Chapter 9 for more information.

1. Square-edged Orifices

$$\dot{M} = (5.983)(K)(d^2)(\text{Fa})(Y)\sqrt{\frac{h_w}{v_1}} \quad (7.9a)$$

2. Flow Nozzles and Venturi Tubes

$$\dot{M} = (5.983)(C)(E)(d^2)(\text{Fa})(\text{Ya})\sqrt{\frac{h_w}{v_1}} \quad (7.9b)$$

WORK

$$\text{Work input, } W = \frac{H}{\eta_p} \frac{\text{Ft-lb}_f}{\text{Lb}_m} \quad (7.10a)$$

or

$$W = (h_2 - h_1)778.16 \quad (7.10b)$$

GAS HORSEPOWER

$$\text{GHP} = \frac{(h_2 - h_1)\,\dot{M}}{42.41} \quad (7.11)$$

or

1. Polytropic

$$\text{GHP}_p = \frac{H_p}{\eta_p} \frac{\dot{M}}{33000} \quad (7.12)$$

2. Adiabatic

$$\text{GHP}_{ad} = \frac{H_{ad} \times Q_1 \times 144 \times P_1}{\eta_{ad} \times T_1 \times R \times 33000} \quad (7.13)$$

$$\text{GHP}_{ad} = \frac{k}{k-1}\left[\left(\frac{P_2}{P_1}\right)^{(k-1)/k} - 1\right] \times \left(\frac{144 \times P_1}{\eta_{ad} \times 33000}\right) \times Q_1 \quad (7.14)$$

SHAFT HORSEPOWER

$$\text{SHP} = \text{GHP} + \text{Bearing and Seal Horsepower} \quad (7.15)$$
$$= \text{GHP} + (\text{GPM} \times \Delta T/12.6), \text{ for light turbine oil}$$

REYNOLDS NUMBER

1. Pipe Flow

$$\text{Re} = \frac{Vd}{\nu'} \quad (7.16)$$

where
V = Gas Velocity
d = Pipe Diameter

2. Compressor Reynolds number

$$\text{Re} = \frac{Ub}{\nu'} \quad (7.17)$$

where
U = Impeller or blade tip velocity, first stage (Ft/Sec)
b = Impeller tip opening, first stage (Ft) centrifugal compressor (axial blade height at impeller outer diameter)
b = Cord length, first stage (Ft), axial compressor

TIP VELOCITY

$$U = N\pi d/720 \quad (7.18)$$
$$= Nd/229.3$$

where
N = Speed
d = Tip Diameter, in.

SPECIFIC VOLUME AND DENSITY

$$v = \frac{1}{\rho} = \frac{ZRT}{144\,P} \quad (7.19)$$

ACOUSTIC VELOCITY

$$a = \sqrt{kg_cRT_1} \quad (7.20)$$

where
k = Ratio of specific heats
g_c = Gravitational constant
R = Gas constant
T_1 = Inlet temperature

MACH NUMBER

$$M = V/a \tag{7.21}$$

VISCOSITY

$$\begin{aligned} \nu' &= (\text{Viscosity in centipoise}) \times 6.72 \times 10^{-4} \times \nu \\ &= (\text{Viscosity in centistokes}) \times 1.076 \times 10^{-5} \\ &= \text{Kinematic viscosity, ft}^2/\text{sec} \end{aligned} \tag{7.22}$$

TOTAL TEMPERATURE

Temperature readings taken during a performance test are usually static readings. This is defined as the temperature that would be shown by a measuring instrument moving at the same velocity as the fluid stream. In making performance calculations it is necessary to consider the effect of velocity on the temperature readings. This new temperature is the stagnation or total temperature and is defined as the temperature which would be measured at the stagnation point if the gas stream were brought to a rest and its kinetic energy converted to an enthalpy rise by an isentropic compression process from the flow condition to the stagnation condition.

Generally it is not necessary to be concerned with converting the total temperature, as the effect of velocity is small. Occasionally, however, readings are taken where velocities are high, such as the return channel readings for sideloads or extraction nozzles. In these cases, total temperature conversion may be required.

Static temperature readings can be converted to total temperature readings using the following equations.

$$T_{\text{total}} = T_{\text{static}} + (1 - .65)\, T_{\text{velocity}}$$

where

$$T_{\text{velocity}} = \frac{V^2}{2g_c 778 c_p} \tag{7.23}$$

where V = Fluid velocity.

Note: The factor 0.65 is an assumed velocity recovery factor. All measurable data are between static temperature and total temperature.

TOTAL PRESSURE

The normal pressure reading taken during a performance test is static pressure. Static pressure is the pressure in the gas measured in such a manner that no effect is produced by the velocity of the gas stream. However, in making performance calculations a different pressure, total or stagnation pressure, is required. This pressure is the pressure which would be measured at the stagnation point when a moving gas stream is brought to rest and its kinetic energy is converted to an enthalpy rise by an isentropic compression from the flow condition to the stagnation condition. As with total temperature, conversion to total pressure is not always required. Only in high velocity conditions, such as internal measurements for sideloads or extractions, is this procedure required.

The conversion from static to total pressure can be made as follows:

$$P_{\text{total}} = P_{\text{static}} + P_{\text{velocity}}$$

where

$$P_{\text{velocity}} = \frac{V^2}{2g_c 144 v} \tag{7.24}$$

where

V = Average fluid velocity = $Q/60A$
Q = Volume flow, CFM
A = Pipe area (FT^2)
v = Local specific volume

ABBREVIATED PARAMETERS

In order to monitor the performance of a compressor, the best thing to do is to monitor head and efficiency versus flow as described in the previous section and compare predicted and previous test data on a continuous day-to-day basis. This will give you the up-to-date performance and trend information needed to predict when a maintenance shutdown is required for performance reasons and/or to help troubleshoot aerodynamic problems. Since a thorough test is not always practical on a continuous basis, some abbreviated parameters or methods of calculation are demonstrated here. When using these methods and procedures, remember that they are approximations and are not meant to replace the performance test described previously, but only to monitor trends.

Pressure Ratio A close look at the equations for calculating head (Equations (7.4) and (7.5)) will show that the primary variable in the equation is pressure, P_2 and P_1. Therefore monitoring pressure ratio will give trends similar to monitoring head. Also, the plot of r_p versus Q will be similar to head versus Q.

$$r_p = \frac{P_2}{P_1} \tag{7.25}$$

Temperature Rise Compressor efficiency is defined as useful work done on the gas (head), divided by the total work. Since total work is directly related to enthalpy, which in turn is related to temperature, monitoring temperature rise will be an indication of total work input. If pressure ratio (r_p) goes down and/or temperature rise goes up (for a given flow and speed), then this is an indication that the efficiency of the compressor has gone down.

$$\Delta T = T_2 - T_1 \tag{7.26}$$

Flow: Orifice and Nozzle Meters At many installations, flow is monitored in percent of some design amount:

$$Q = \left(\frac{\%}{100}\right) \times Q_D \tag{7.27}$$

where

Q_d = 100% design meter flow rate, SCFM
% = Meter reading, %

Since this 100% flow was calculated for the design condition of the flow meter, the flow must be corrected for the actual conditions at the flow meter:

$$Q_c = Q \times \sqrt{\frac{P_A}{P_D} \times \frac{T_D}{T_A} \times \frac{MW_A}{MW_D} \times \frac{Z_D}{Z_A}} \tag{7.28}$$

where

Q_c = Corrected flow, SCFM
A = Actual condition at flow meter
D = Design condition of flow meter.

Also, the compressor characteristics are dependent on actual inlet flow, so the standardized flow must be corrected to inlet conditions.

$$Q_1 = Q_c \times \frac{P_S}{P_1} \times \frac{T_1}{T_S} \times \frac{Z_1}{Z_S} \tag{7.29}$$

where

S = Standard conditions
1 = Inlet conditions

Flow: Casing Nozzle Meter Some equipment has built-in flow meters. Such is the case with the Elliott axial compressor. The flow in most compressors is generally accelerated somewhat before entering the first stage of compression. This can be equated to a Venturi effect and used to monitor flow rate.

In the case of the Elliott axial, each compressor inlet is calibrated during performance testing. The results are then plotted for various suction temperatures. (See Figure 7.9.)

TREND ANALYSIS

Accurate trend analysis on compressors can be very difficult and confusing as the operating point and even the gas analysis continuously change. Since this alone will affect the efficiency of the compressor, how can the trend be evaluated?

One method that has been used with some success is to plot percentage change of a parameter to a known baseline (see Figure 7.6).

It is important to first establish the base line. Generally, the predicted performance curve is used. Preferably, the predicted curve is adjusted according to established field data for the compressor. Adjustments must be made for changes in inlet conditions, gas analysis, and speed. The test data are then compared to this "adjusted" prediction curve and the percent difference plotted versus time. Since performance degradation can be greater for off-design conditions, it is necessary to compensate for this. Some "educated guessing" is then required to determine how much the performance curve has actually shifted. As shown in Figure 7.6, the data can be very deceiving, and therefore caution must be used in interpreting the data. Actual operating range will determine the urgency of any maintenance shut down.

ENHANCED TREND METHOD

When compressor performance decays due to polymer buildup, dirt, corrosion, increased internal recirculation from seal wear, etc., the performance curve generally shifts downward and toward reduced flow as shown in Figure 7.7.

Additionally, the process system may also be fouling. This means a greater restriction for a given flow. So, while the compressor has less head capability, more head may be required by the system.

The efficiency is reduced because of the increased frictional losses and/or increased internal recirculation, shifting the curves down. This increased resistance also effectively reduces the capacity of the compressor, shifting the curves to the left. Even the shape of the curve will change some.

The problem in trending data can be minimized if a full range of data points is available to plot the full curve. Too often, however, due to production or process restrictions, only one flow point is available at any one point in time. If a test is conducted at point "A" on Figure 7.7, then several months later a test is conducted at data point "B," it appears, at least on the surface, that performance has degraded badly. However if it were possible to take a full range of data it would be found that both points A and B represent the same degree of fouling. The trend is flat.

One way of compensating for all of this is demonstrated in Figure 7.8. It is feasible to have the equipment manufacturer analytically predict the relative effect of fouling. This can be done for the overall compressor or for just the last stage. By plotting several degrees of fouling on a speed-compensated plot, it is easy to distinguish the trend of the fouling. The "degree of fouling" is plotted versus time with reasonable confidence and prediction of timing for maintenance is simplified.

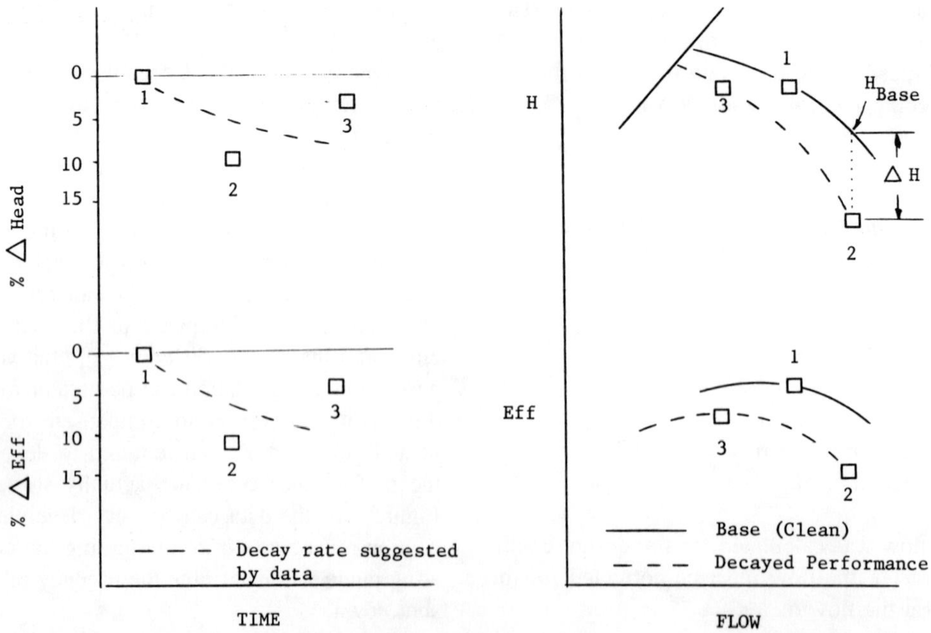

$$\% \Delta H = \left(\frac{H_{Base} - H_2}{H_{Base}} \right) x\ 100\%$$

Figure 7.6. Trend analysis of a compressor performance. Note how the high flow rate data (Point 2) suggest that the trend of decay in performance is rapid while data taken later at a lower flow rate (Point 3) show the decay to be at a lesser rate. Actually, there has been no decay in performance from Point 2 to Point 3.

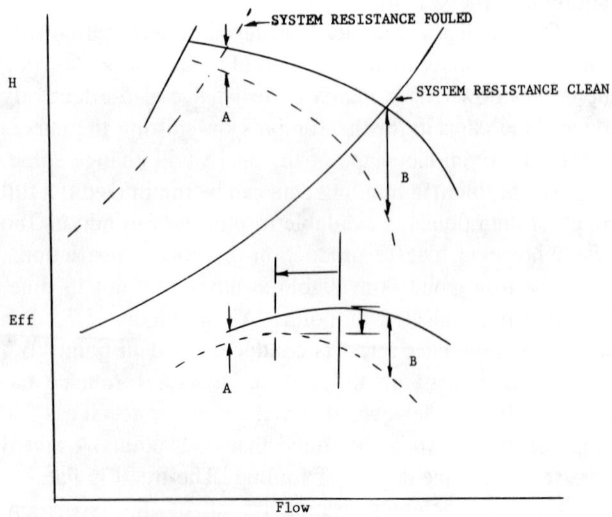

Figure 7.7. Effect of fouling on compressor performance curves.

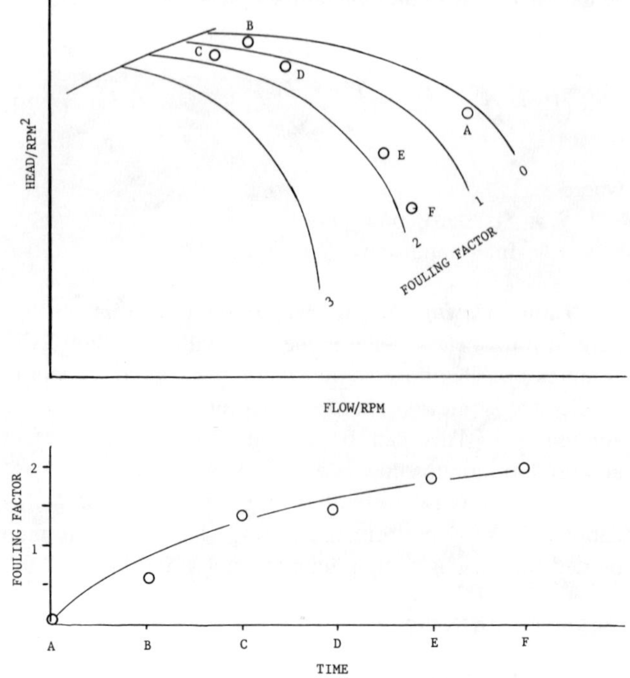

Figure 7.8. Plot performance data by using a fouling factor curve.

SAMPLE PROBLEM 7.1
Using Gas Properties Procedure
Gas analysis:

H_2	Hydrogen	34.12%
C_2H_4	Ethylene	33.57%
C_2H_6	Ethane	23.86%
CH_4	Methane	5.07%
CO	Carbon Monoxide	3.38%

$P_1 = 233.4$ psig $t_1 = 91.4°F$
$P_2 = 512.2$ psig $t_2 = 257°F$

Speed = 6350 RPM

Atmospheric Pressure = 14.5 psi

First convert data to the proper units.

$P_1 = 233.4 \text{ psig } + 14.5 = 247.9 \text{ psia}$
$P_2 = 512.2 \text{ psig } + 14.5 = 526.7 \text{ psia}$
$T_1 = 91.4°F + 460 = 551.4°R$ 273
$T_2 = 257°F + 460 = 717°R$

From Gas Properties Program

Inlet Conditions

$P = 247.9$	$h = 220.3$
$Z = .9665$	$v = 1.212$
$T = 551.4$	$k = 1.265$
$MW = 19.04$	$c_p = .537$

Discharge Conditions

$P = 526.7$	$h = 309.0$
$Z = .9803$	$v = .7524$
$T = 717.0$	$k = 1.240$
$MW = 19.04$	$c_p = .609$

Flow Meter Conditions

$P = 249.9$	$h = 220.4$	$\nu' = .01048$
$Z = .9664$	$v = 1.203$	
$T = 551.7$	$k = 1.265$	
$MW = 19.04$	$c_p = .538$	

Flow Rate

Pipe Diameter = 16″ OD, 15.25″ ID
Orifice Diameter = 9.15″
$\beta = 9.15/15.25 = .6$
Orifice ΔP = 128.9 in water

Calculate Reynolds number

$$Re = \frac{Vd}{\nu'}$$

where

$$\begin{aligned} \nu' &= \text{Kinematic viscosity ft}^2/\text{sec} \\ &= \text{Centipoise } \times 6.72 \times 10^{-4} \times \text{specific volume} \\ &= 0.01048 \times 6.72 \times 10^{-4} + 1.203 \\ &= 8.47 \times 10^{-6} \end{aligned} \tag{7.22}$$

In order to calculate Reynolds number, the flow rate must first be known. For a start, use the design flow rate for the compressor.

Assume $Q = 3600$ CFM:

$$V = \frac{Q}{A} = \frac{3600 \text{ ft}^3}{\text{min}} \times \frac{4}{(3.14)(15.25 \text{ in})^2} \times \frac{\text{min}}{60 \text{ sec}} \times \frac{144 \text{ in}^2}{\text{ft}^2}$$

$$= 47 \text{ ft/sec}$$

$$Re = \frac{47 \text{ ft}}{\text{sec}} \times \frac{\text{sec}}{8.47 \times 10^{-6} \text{ ft}^2} \times 15.25 \text{ in} \times \frac{\text{ft}}{12 \text{ in}}$$

$$= 7.05 \times 10^6$$

From Chapter 9, Equation (9.1):

$$\dot{M} = (5.983)(K)(d^2)(\text{Fa})(Y)\sqrt{\frac{h_w}{v_1}}$$

$$K = \frac{C}{\sqrt{1-\beta^4}}$$

$$\dot{M} = 5.982 \frac{CYd^2\text{Fa}}{\sqrt{1-\beta^4}} \sqrt{\frac{h_w}{v_1}}$$

$C = 0.6049$ (from Table 9.2)

Fa $= 1.000$ (from Figure 9.3)

$k = 1.265$

$x = \Delta P/P_1 = 128.9 \times .03613/250 = .0186$

$x/k = .0186/1.265 = .015$

$Y = .993$ from Figure 9.4

$$\dot{M} = 5.982\left(\frac{0.6049 \times 0.993 \times 9.15^2 \times 1.00}{\sqrt{1-.60^4}}\right)\sqrt{\frac{128.9}{1.203}}$$

$$= 3337 \text{ lb/min}$$

At compressor inlet:

$$Q = \dot{M} \times v_1$$
$$= 3337 \frac{\text{lb}}{\text{min}} \times 1.212 \frac{\text{ft}^3}{\text{lb}}$$
$$= 4044 \text{ ft}^3/\text{min}$$

Calculate Head

$$H = Z_1 R T_1 \left(\frac{n}{n-1}\right)[(r_p)^{(n-1)/n} - 1] \quad (7.4a)$$

$$Z_1 = 0.966$$

$$R = 1545/\text{MW} = 1545/19.04 = 81.14$$

$$n = \frac{\ln (P_2/P_1)}{\ln (v_2/v_1)}$$

$$v_1 = 1.212 \text{ ft}^3/\text{lb}$$

$$v_2 = 0.7524 \text{ ft}^3/\text{lb}$$

$$n = \frac{\ln (526.7/247.9)}{\ln (1.212/0.7524)} = \frac{0.7536}{0.4768} = 1.581$$

$$r_p = \frac{P_2}{P_1} = \frac{526.7}{247.9} = 2.125$$

$$H = (0.966)(81.14)(551.4)\left(\frac{1.581}{0.581}\right)[2.125^{(.581/1.581)} - 1]$$

$$H = 37538 \text{ ft.}$$

Polytropic Efficiency

$$\eta_p = \frac{H_p}{(h_2 - h_1)778.16} \quad (7.6)$$
$$= \frac{37538}{(309 - 220.3)778.16}$$
$$= 0.545$$

Gas Horsepower

$$\text{GHP} = \frac{(h_2 - h_1)\dot{M}}{42.41} \quad (7.11)$$
$$= \frac{(309 - 220.3)3337}{42.41}$$
$$= 6979 \text{ HP}$$

Power Balance

Motor Power

$$\text{BHP} = E \times I \times \eta \times \text{P.F.} \times \sqrt{3}/746$$
$$= 6600 \times 444 \times .96 \times .92 \times 1.732/746$$
$$= 6009 \text{ HP}$$

Mechanical Losses

Mechanical losses are generally a small percentage of the overall horsepower, therefore exact numbers are not critical to the overall outcome. If the values are significant, then the losses can be verified by measuring oil flow rates and temperature rise. For light turbine oil (32 SSU), HP = oil flow rate in gpm $\times \Delta T/12.6$.

57 compressor bearings and seals
21 motor bearings
85 gear
163 total

$$\text{Test Error} = \left[\left(\frac{\text{Driver Power} - \text{Losses}}{\text{GHP}}\right) - 1\right] \times 100\%$$
$$= \left[\left(\frac{6009 - 163}{6979}\right) - 1\right] \times 100\%$$
$$= 16.2\% \text{ Error}$$

Note that compressor power is higher than driver power and efficiency is low! This indicates something is probably wrong with the performance test. Recheck driver power, gas properties, pressure measurement method, etc.

EXAMPLE PERFORMANCE TEST 7.2
Axial Air Compressor
Test Data

P_1	= 14.5 psia	T_1	= 56°F = 516°R
P_2	= 54.5 psia	T_2	= 349°F = 809°R
Inlet ΔP	= 40 in. H_2O	N	= 5804 RPM

Head

$$H_{ad} = RT_1\left(\frac{k}{k-1}\right)\left[\left(\frac{P_2}{P_1}\right)^{(k-1)/k} - 1\right] \quad (7.5a)$$
$$= \left(\frac{1545}{28.97}\right) \times 516\left(\frac{1.4}{1.4-1}\right)\left[\left(\frac{54.5}{14.5}\right)^{(1.4-1)/1.4} - 1\right]$$

$= 53.33 \times 516 \times 3.5(3.76^{.286} - 1)$
$= 44353$ ft-lb/lb

Efficiency

$$\eta_{ad} = \frac{T_1[(P_2/P_1)^{(k-1)/k} - 1]}{T_2 - T_1}$$

$$= \frac{516[(54.5/14.5)^{(1.4-1)/1.4} - 1]}{809 - 516} \qquad (7.8)$$

$$= \frac{516 \times .46}{293}$$

$$= .81$$

Horsepower

Inlet flow for this compressor is easy to check since it has a calibrated inlet. From calibration curves, Figure 7.9,

$Q_1 = 92000$ ICFM

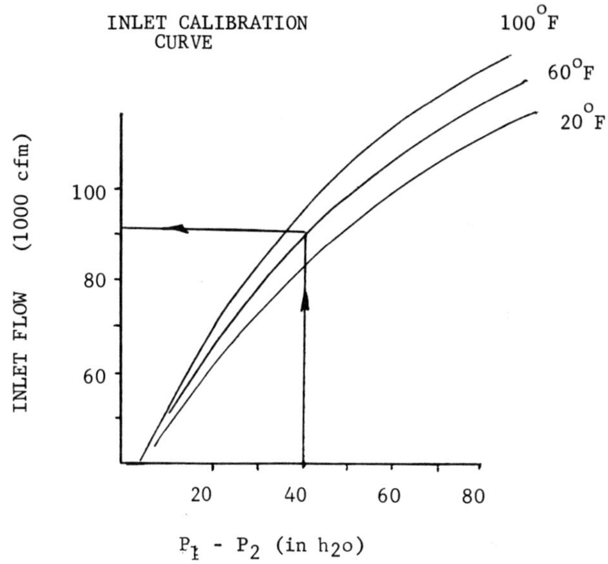

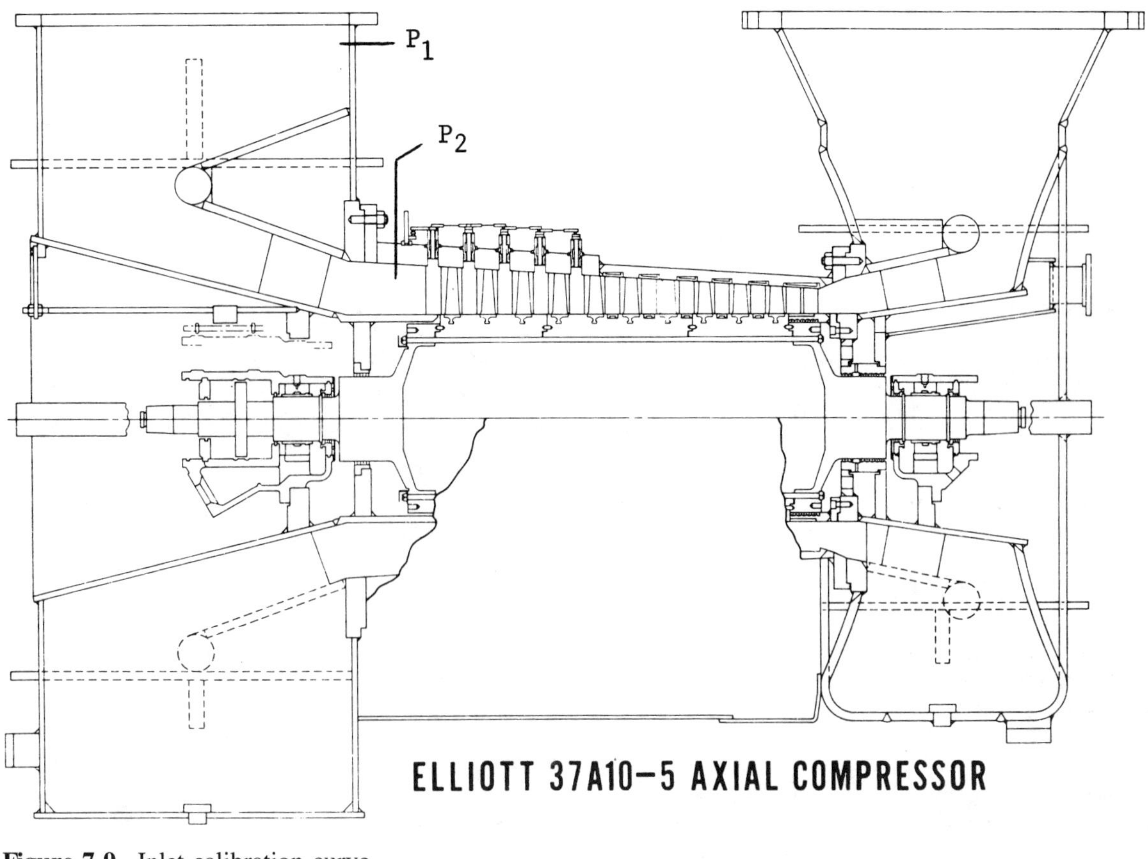

Figure 7.9. Inlet calibration curve.

$$\text{GHP} = \frac{H_{ad} \times Q \times 144 \times P_1}{\eta_{ad} \times T_1 \times R \times 33000}$$

$$= \frac{44353 \times 92000 \times 144 \times 14.5}{.81 \times 516 \times 53.33 \times 33000} \quad (7.13)$$

$$= 11583$$

Power Balance

Motor Power

$$\text{BHP} = E \times I \times \eta \times \text{P.F.} \times 1.732/746$$
$$= 13800 \times 442 \times .95 \times .91 \times 1.732/746$$
$$= 12244 \text{ HP}$$

Mechanical Losses

75 compressor bearings
35 motor bearings
235 gear
345 total

$$\text{Test Error} = \left[\left(\frac{\text{Driver Power} - \text{Losses}}{\text{GHP}}\right) - 1\right] \times 100\%$$

$$= \left[\left(\frac{12244 - 345}{11588}\right) - 1\right] 100\%$$

$$= 2.7\%$$

EXAMPLE 7.3
Mollier Method
(Note that it is possible to obtain Mollier Diagrams for any gas or gas mixture.)

100% Propane

$P_1 = 14.5$ psia $T_1 = 40°\text{F}, 500°\text{R}$
$P_2 = 305$ psia $T_2 = 312°\text{F}, 772°\text{R}$

Flow Meter, 85%

Flow Meter Design Conditions

$Q = 6.0$ MMSCFD (million standard cubic feet per day)
$P = 16.7$ psia MW = 44
$T = 50°\text{F}$ $Z = .977$

Gas Properties from Mollier Diagram, Figure 7.10

$v_1 = 8.4\ \text{ft}^3/\text{lb}$ $v_2 = .57\ \text{ft}^3/\text{lb}$
$h_1 = 128$ BTU/lb $h_2 = 242$ BTU/lb

$$Pv = ZRT$$

$$Z = \frac{Pv}{RT} \quad R = \frac{1545}{44} = 35\frac{\text{ft-lb}_f}{\text{lb}_m\text{-°R}}$$

$$Z_1 = \frac{14.5 \times 8.4 \times 144}{35 \times 500}$$

$$= 1.00$$

Flow Rate

$$Q = \frac{85\%}{100} \times 6.0 \text{ MMSCFD}$$

$$= 5.1 \text{ MMSCFD}$$

$$= \frac{5.1 \text{ MM STD.FT}^3}{\text{DAY}} \times \frac{10^6 \text{ STD. FT}^3}{\text{MM STD. FT}^3} \times \frac{\text{DAY}}{24 \text{ HRS}} \times \frac{\text{HR}}{60 \text{ MIN}}$$

$$= 3542 \text{ SCFM (standard cubic feet per minute)}$$

Using Eq. (7.28):

$$Q_c = Q\sqrt{\frac{P_A}{P_D} \times \frac{T_D}{T_A} \times \frac{\text{MW}_A}{\text{MW}_D} \times \frac{Z_D}{Z_A}}$$

$$= 3542\sqrt{\frac{14.5}{16.7} \times \frac{510}{500} \times 1.0 \times \frac{.977}{1.0}}$$

$$= 3294 \text{ SCFM}$$

Using Eq. (7.29):

$$Q_1 = Q_c \times \frac{P_S}{P_1} \times \frac{T_1}{T_S} \times \frac{Z_1}{Z_S}$$

$$= 3294\left(\frac{14.7}{14.5} \times \frac{500}{520} \times \frac{1.0}{1.0}\right)$$

$$= 3212 \text{ ICFM (inlet cubic feet per minute)}$$

Head

$$H_p = 72\left[\ln\left(\frac{P_2}{P_1}\right)\right](P_1 v_1 + P_2 v_2) \quad (2.36)$$

English units

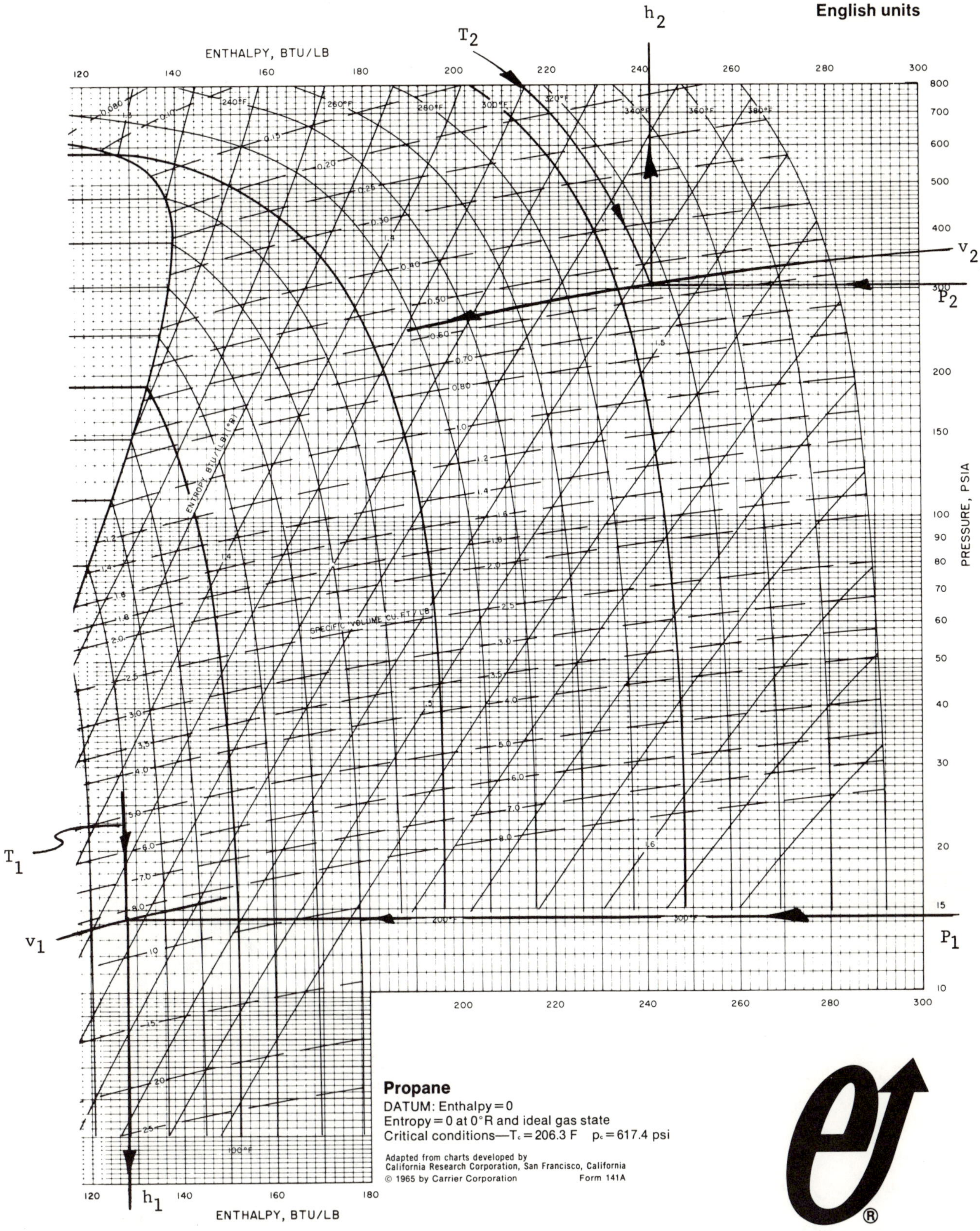

Figure 7.10. Mollier diagram for propane (see Example 7.3).

$$= 72\left[\ln\left(\frac{305}{14.5}\right)\right](14.5 \times 8.4 + 305 \times .57)$$

$$= 64800 \qquad \text{(Ref. 63966)*}$$

Polytropic Efficiency

$$\eta_p = \frac{H_P}{(h_2 - h_1)778.16} \qquad (7.6)$$

$$= \frac{64800}{(242 - 128)778.16}$$

$$= .73 \qquad \text{(Ref. .726)}$$

Gas Horsepower

$$\dot{M} = \frac{Q}{v}$$

$$= \frac{3212}{8.4}$$

$$= 382 \text{ \#/min}$$

$$\text{GHP} = \frac{H_p \dot{M}}{\eta_p\ 33000} \qquad (7.12)$$

$$= \frac{64800 \times 382}{.73 \times 33000}$$

$$= 1027 \text{ HP}$$

Power Balance

Motor Power

$$\text{HP} = E \times I \times \eta \times \text{P.F.} \times 1.732/746$$
$$= 6600 \times 85 \times .95 \times .91 \times 1.732/746$$
$$= 1125$$

Mechanical Losses

45 compressor bearings and seals
12 motor bearings
22 gear
79 total

$$\text{Test Error} = \left[\left(\frac{\text{Driver Power} - \text{Losses}}{\text{GHP}}\right) - 1\right] \times 100\%$$

$$= \left[\left(\frac{1125 - 79}{1027}\right) - 1\right] \times 100\%$$

$$= 1.9\%$$

EXAMPLE 7.4
Hand Calculate Gas Properties [11].
Gas Analysis

Propane	.89
Butane	.06
Ethane	.05

$P_1 = 20$ psia $\quad T_1 = 40°F = 500°R$
$P_2 = 100$ psia $\quad T_2 = 180.5°F = 640.5°R$
$N = 10650$ RPM
$Q = 5280$ ICFM

Use Table 7.1 to calculate

- MW of mixture
- k of mixture
- Z of mixture

Find Z_1 using Figure A.1. First find P_{R_1} and T_{R_1}:

$$P_{R_1} = P_1/P_c, \quad T_{R_1} = T_1/T_c$$

$$P_{R_1} = \frac{20}{617.} = .0324 \quad T_{R_1} = \frac{40 + 460}{666.} = .75$$

$$Z_1 = .97 \text{ (from Appendix A, Figure A.1)} \qquad \text{(Ref. .970)}$$

$$v_1 = ZRT/144P \qquad (7.19)$$

$$v_1 = .97 \times \frac{1545}{44.23} \times \frac{(40 + 460)}{144 \times 20}$$

$$= 5.88 \text{ ft}^3/\text{lb} \qquad \text{(Ref. 5.884)}$$

$$P_{R_2} = \frac{P_2}{P_c} = \frac{100}{617.} = 0.162$$

*Reference numbers indicate values obtained using the gas properties procedure demonstrated in Sample Problem 7.1.

TABLE 7.1 Calculation of Properties for a Gas Mixture

Gas Mixture	(1) Mol% each gas	(2) Mols/hr (Mol% × 2400)	(3) Mol. Mass	(4) (1) × (3)	(5) Mass% [(4) ÷ 44.23] × 100	(6) T_c,°R	(7) p_c, psi	(8) (1) × (6)	(9) (1) × (7)	(10) $C_{p,m}$ Btu/mol-F	(11) (1) × (10)
Propane	89%	2136	44.09	39.24	88.72%	666	617	592.7	549.1	16.58	14.76
n-Butane	6%	144	58.12	3.49	7.89%	766	551	46.0	33.1	22.53	1.35
Ethane	5%	120	30.07	1.50	3.39%	550	708	27.5	35.4	11.98	0.60
		2400		44.23				666.2	617.6		16.71
				Apparent Mol. Mass of Mixture				T_c(mix)	p_c(mix)		$C_{p,m}$(mix)

$$k(\text{Mixture}) = \frac{16.71}{16.71 - 1.99} = 1.135$$

Note: Items 3, 6, 7, and 10 are obtained from Table A.1 in Appendix A.

$$T_{R_2} = \frac{T_2}{T_c} = \frac{640.5}{666.} = .961$$

From Figure A.1, $Z_2 = 0.93$ (Ref. .9323)

$$v_2 = .93 \times \frac{1545}{44.23} \times \frac{640.5}{144 \times 100}$$

$= 1.44$ ft^3/lb (Ref. 1.449)

Head

$$H_P = 72\left[\ln\left(\frac{P_2}{P_1}\right)\right](P_1 v_1 + P_2 v_2) \quad (2.36)$$

$$= 72\left[\ln\left(\frac{100}{20}\right)\right](20 \times 5.88 + 100 \times 1.44)$$

$= 30300$ (Ref. 30578)

Efficiency

$$\eta_P = \left(\frac{k-1}{k}\right) \div \left(\frac{n-1}{n}\right) \quad (7.7)$$

where

$$n = \frac{\ln(P_2/P_1)}{\ln(v_1/v_2)} = \frac{\ln(100/20)}{\ln(5.88/1.144)} = 1.144$$

$$\eta_p = \left(\frac{1.135-1}{1.135}\right) \div \left(\frac{1.144-1}{1.144}\right) = 0.95$$

(Ref. .716)

Obviously this is an incorrect answer. The compressor cannot possibly have an efficiency of 95%! So we should go back and recalculate k using an average temperature.

$$T_{\text{average}} = \frac{T_1 + T_2}{2}$$

$$= \frac{40 + 180.5}{2}$$

$= 110$°F

From Table A.1 for propane:

$Cp,m^* = 16.82$@ 50°F, 23.57 @ 300°F

Interpolate for value @110°F

$$\frac{110 - 50}{300 - 50} = \frac{X - 16.8}{23.57 - 16.8}$$

$$\frac{60}{250} = \frac{X - 16.8}{6.77}$$

$$X = 1.63 + 16.8$$

Cp,m @110°F $= X = 18.4$

Values can be obtained for butane and ethane in a similar fashion.

Butane
$Cp,m = 24.81$ @110°F

Ethane
$Cp,m = 13.14$ @110°F

*Cp,m and Mc_p are used interchangeably. See "Gas Mixtures" in Chapter 2.

To find k at the average temperature, first find Cp,m (mix) using the Cp,m values at average temperature just calculated.

$$
\begin{aligned}
Cp,m(\text{mix}) &= .89 \times 18.4 + .06 \times 24.81 + .05 \times 13.14 \\
&= 16.38 + 1.49 + .67 \\
&= 18.52
\end{aligned}
$$

$$\text{mixture} = \frac{18.52}{18.52 - 1.99} \qquad (2.54)$$

$$= 1.12$$

Using this new value of k for average conditions calculate Eff.

$$\eta_p = \frac{(k-1)/k}{(n-1)/n} \qquad (7.7)$$

$$= \frac{(1.12-1)/1.12}{(1.144-1)/1.144}$$

$$= .85 \qquad (\text{Ref. } .716)$$

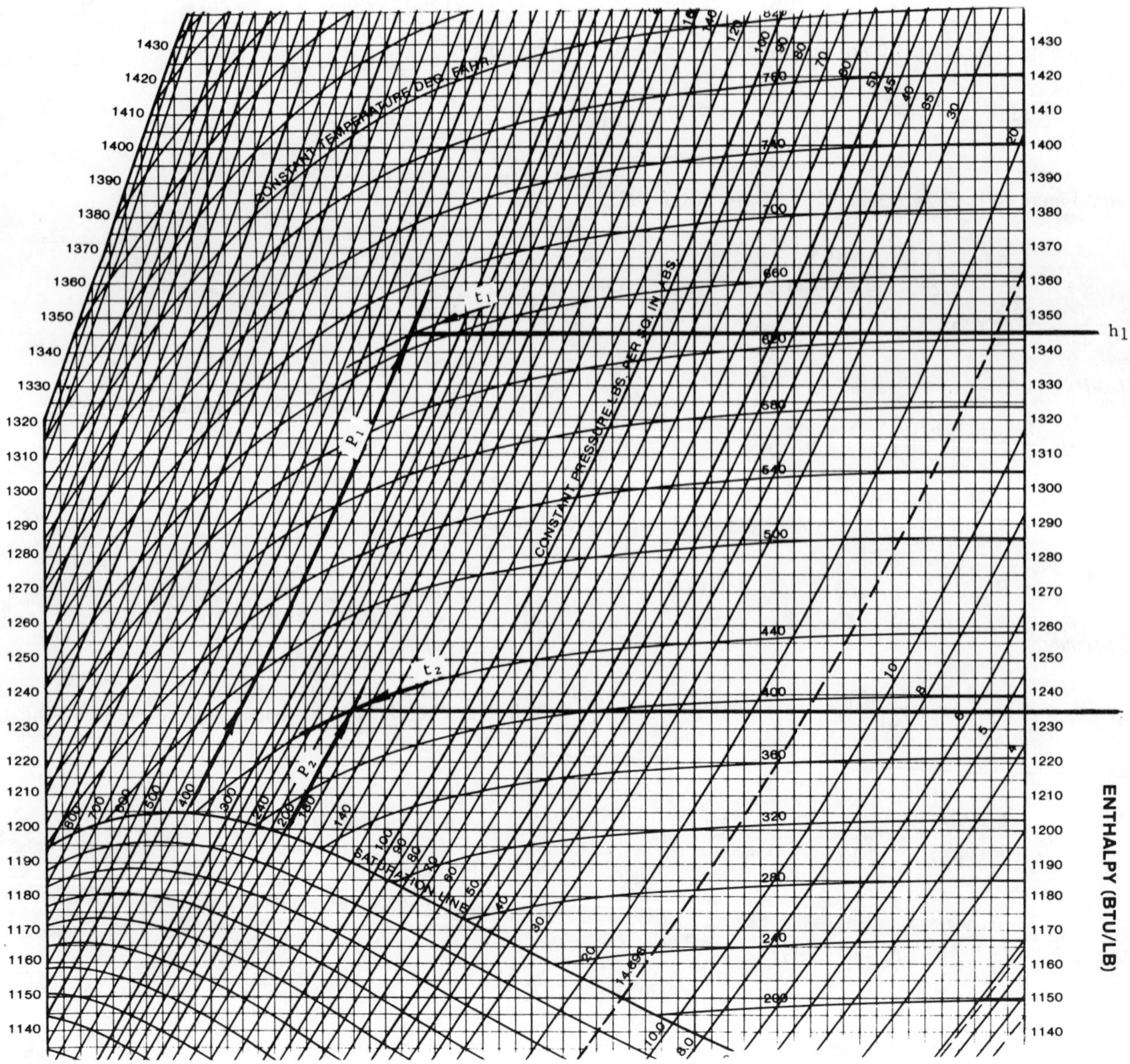

Figure 7.11. Mollier diagram for steam (Example 7.3).

This is still an inaccurate evaluation of the compressor efficiency. (Note the reference value of .716 evaluated by a gas properties program.)

The efficiency cannot accurately be hand-calculated for this problem. This is common for high mole weight gases. The problem is due to the nonlinear relationship of the gas properties near the dew point. When looking at values far from the dew point, such as with air or nitrogen, the values are near linear and perfect gas laws are accurate. See the Mollier diagrams in the reference section and the situation will be apparent.

Power

Since efficiency cannot accurately be established, then also work and power cannot be established. We can, however, work backward from the driver to establish the gas power.

Driver—Steam Turbine (from Figure 7.11):

$$t_1 = 670°\text{F} \qquad P_1 = 400 \text{ psia}$$
$$t_2 = 527°\text{F} \qquad P_2 = 200 \text{ psia}$$
$$h_1 = 1345 \qquad h_2 = 1235$$

Steam flow = 506 #/min

$$\begin{aligned} \text{GHP} &= \dot{M}(h_1 - h_2)/42.4 \\ &= 506(1345 - 1235)/42.4 \qquad (7.11) \\ &= 1313 \text{ HP} \end{aligned}$$

Mechanical Losses

(Estimated Losses)

65 compressor
53 turbine

118 total

$$\text{GHP} = 1313 - 118 = 1195 \text{ HP}$$

(Compressor gas horsepower)

Compressor Efficiency (See Equations (2.45) through (2.47))

$$\text{Efficiency} = \frac{\text{Head}}{\text{Work}}$$

$$\text{Work} = \frac{\text{GHP}}{\dot{M}}$$

$$\dot{M} = Q/v_1$$

$$= \frac{5280 \text{ ft}^3}{\text{min}} \times \frac{\text{lb}}{5.88 \text{ ft}^3}$$

$$= 898 \text{ lb/min}$$

$$\text{Work} = 1195 \text{ HP} \times \frac{\text{min}}{898 \text{ lb}_\text{m}} \times \frac{33000 \text{ ft-lb}_\text{m}}{\text{HP-min}}$$

$$= 43900 \, \frac{\text{ft-lb}_\text{f}}{\text{lb}_\text{m}}$$

$$\eta = \frac{30300}{43900} = .69 \qquad \text{(Ref. .716)}$$

Now plot data on non-dimensionalized curves to minimize effect of small speed changes. (See Figure 7.12.)

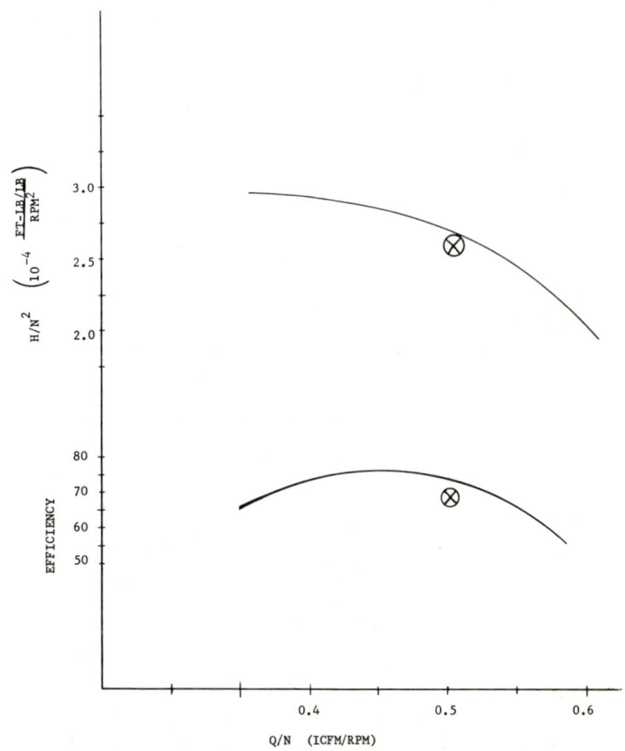

Figure 7.12. Nondimensionalized performance curve (Example 7.4).

$$H/N^2 = 30300/10650^2 = 2.67 \times 10^{-4}$$

$$Q/N = 5280/10650 = 0.498$$

Plotting the data on the curves shows that both efficiency and head are lower than but close to predicted values, therefore the test was probably of reasonable accuracy.

Note: An alternate way to measure horsepower is to use a coupling equipped with a torque meter. Some useful equations are:

$$\text{HP} = 550 \text{ ft-lb/sec} = 33000 \text{ ft-lb/min}$$

$$= \frac{\text{torque} \times \text{speed} \times 2\pi}{33000}$$

$$= \text{torque} \times \text{speed}/5252$$

where

Torque = ft-lb
Speed = rpm

From the torque meter reading between the turbine and compressor:

$$\text{Torque} = 605 \text{ ft-lb}$$

$$\text{SHP} = \frac{605 \times 10650}{5252}$$

$$= 1230 \text{ HP}$$

Power Balance

From Torque meter:

$$\text{SHP} = 1230$$
$$\text{GHP} = 1230 - 65$$
$$= 1165 \text{ HP}$$

From steam tables:

$$\text{GHP} = 1195$$

$$\text{Test error} = \left(\frac{1195}{1165} - 1\right) 100\%$$

$$= 2.6\%$$

8

TROUBLESHOOTING

Table 8. 1 is a guideline to help troubleshoot aerodynamic-related problems. As with any troubleshooting situation, there are some things that need to be done first.

1. *Define the problem.*
   a. *What exactly is the problem?*
   b. *What should the performance be?*
   c. *What is the performance now?*
2. *Outline the history of the compressor.*
   a. *How long has it been operating?*
   b. *When was the last overhaul?*
   c. *What changes were made at that time?*
   d. *When did the problem start?*
   e. *Was it a quick or gradual change?*
   f. *Note the trend of various parameters.*
   g. *What else changed, what other problems occurred at this time*
      i. *on the compressor?*
      ii. *in the process?*
      iii. *in operation and control?*
3. *Verify all data.*
   a. *Have instruments been calibrated?*
   b. *Do cross checks agree?*

If a thorough performance test has not been completed, do it now. Get help if it is needed. Be sure to follow as closely as possible the test procedure outlined in the previous chapter. If possible get several operating points at one speed so a full curve can be plotted. This can be a big help in determining corrective measures.

TABLE 8.1 Troubleshooting Checklist

- Define problem—what, where, when
- Outline history of operation—trend data
- Verify data

Test accuracy

- Complete power balance
- Check pressure taps: location, size, condition
- Liquid in pressure lines
- Temperature probe insertion depth, heat transfer
- Calibrate instruments
- Inspect flow meter: wear, sludge buildup
- Condensates in gas analysis
- Vortex or undeveloped velocity profile upstream of flow meter
- Conduct a mass flow balance

Equipment problems

- Vortex or undeveloped velocity profile upstream of compressor suction
- Internal leakage across diaphragm splitline
- Recirculation from rubbed interstage seals or balance piston seals, casing drains, other areas
- Foreign object damage or blockage
- Liquids in process
- Dirt accumulation or polymer buildup
- Erosion of impeller blades and diffuser passages
- Proper direction of rotation
- Balance line sleeve

Economics

- Per diem cost to operate as is
- Associated risks
- Cost for repairs
- Cost for down time to complete repairs
- Safety concerns

COMMON SOURCES OF TEST ERROR

In order to troubleshoot any problem, it is important to have correct information on the subject. Aerodynamic performance is very involved and data errors can rapidly mushroom, thereby misleading the problem solver. It is therefore essential that accurate data be obtained.

Before troubleshooting the compressor, troubleshoot the testing procedure. The best way to do this is a power balance. If it is not feasible to do a power balance, or if there is a significant error (7%) between the compressor power and the driver power, a thorough analysis of the test procedure is necessary. Plotting the compressor work (Equation 7.10) may be helpful.

GAS ANALYSIS

To have good test results it is critical to have an accurate gas analysis. This can be a bit complex on high pressure, high mole weight gas. If the sample is taken at high temperatures, some of the heavy gas may condense on the walls of the sample container when it cools. If you take the sample at the inlet, there may be some liquids in the gas stream that will remain in the sample container. When you test the gas, this condensed liquid will remain in the bottle unless heated.

For best results, take samples at both the inlet and discharge points. Check for condensibles and compensate by heating the sample before testing.

LIQUIDS IN THE SYSTEM

If there is liquid anywhere in the system, it is possible that some may carry over into the compressor. Knockout drums and demister pads do not always work the way they

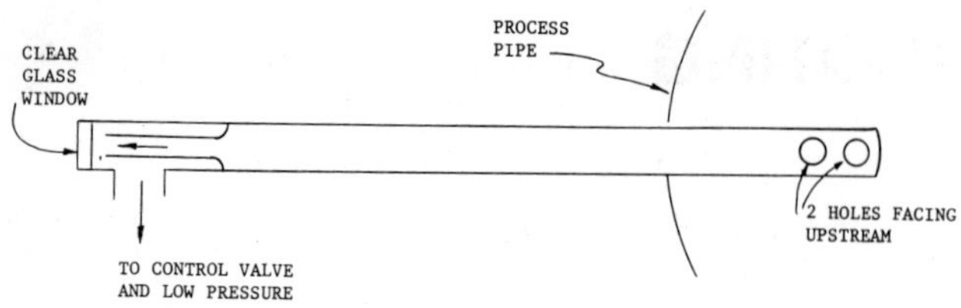

Figure 8.1. Instrument for detecting liquids in process piping. Liquid will cause the viewing window to fog. *(Data from [26].)*

should. This liquid carryover will give erroneous results on the performance test. One way to be certain there is no liquid is to measure the flow rate at each flange, because liquids in the gas stream will also give an erroneous reading to the flow meters. This mass flow balance is a good cross check for both flow measurement accuracy and/or possible liquid ingestion. An instrument for checking for liquid is shown in Figure 8.1.

Another liquid problem is liquid in pressure tap lines. Be sure all lines are properly sloped and drained. If lines are too small (less than 1/2 inch), capillary action will hold liquid in the lines.

Be sure to open drain valves at low spots in process piping and instrument lines before, during, and after test.

PRESSURE AND TEMPERATURE MEASUREMENT

Be sure that a proper pressure tap is installed (see Figure 7.2). Inspect the inside edge of the hole to see that it was deburred and that it has not been eroded, corroded, or plugged with dirt.

Check thermocouple installations. Thermocouples should be inserted into the pipe 1/3 to 1/2 the pipe diameter. Also, it is critical that the thermocouple have intimate contact with the thermowell. Use graphite paste as suggested in Chapter 7.

Although not normally required, the proper method of analyzing compressor performance utilizes "total" pressure and temperature rather than the "static" values that are measured (see Equations (7.23) and (7.24)). Total values

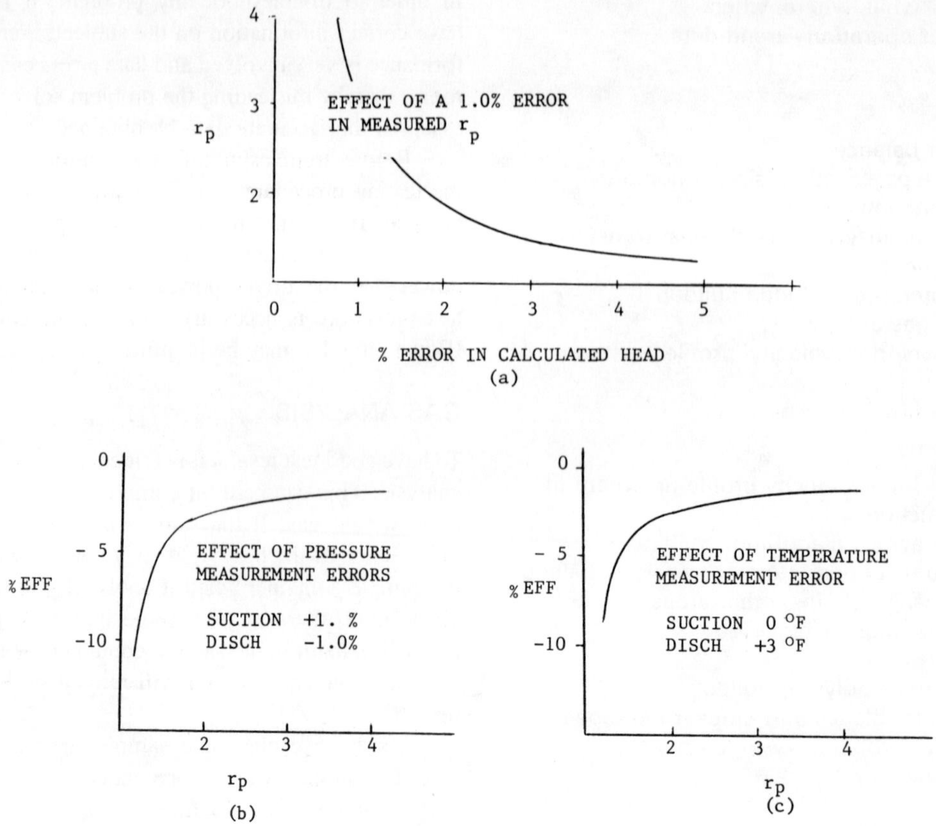

Figure 8.2. Effects of measurement errors on test results. Data for natural gas (based on information from Lock [26]): (a) Effect of a 1% error in measured pressure/ratio on calculated isentropic head; (b) For the suction and discharge measurement errors shown, the percentage point deviation in efficiency is plotted vs. pressure ratio; (c) Percentage point deviation in efficiency for the temperature measurement errors shown.

are usually only required when very high velocities are encountered.

Be sure to use only instruments that have recently been calibrated. Significant errors can be introduced by normal vibration and handling during operation.

As a reference, Figure 8.2 is provided to demonstrate the effect of pressure and temperature error on performance test results.

VELOCITY PROFILE

If possible, a flow meter should be installed in each inlet and discharge pipe so a mass flow balance in the system can be carried out. This is done by simply comparing the total mass inflow to the total mass outflow. The difference is the accuracy of the flow meters.

A major source of problems both in compressor performance and in obtaining accurate flow data is an incomplete velocity profile and/or a vortex upstream of the compressor suction or flow meter. Either situation can seriously alter the compressor and/or flow meter performance.

A flow straightener device is required when flow swirl or a vortex is present. This can occur when there are two or more adjacent elbows in different planes. A flow straightener can be a tube bundle or an "egg crate," as shown in Figure 8.3.

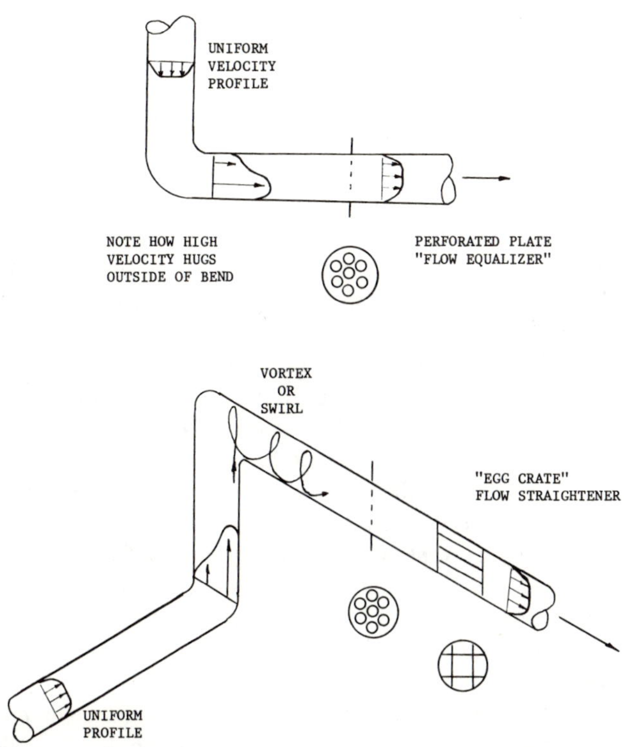

Figure 8.3. Vortex and non-symmetric velocity profile caused by piping arrangement.

A flow equalizer is required when the velocity profile is not uniform. This can be caused by flow hugging one side of a pipe due to flow around an elbow or flow through a partly closed butterfly valve. This situation is best corrected by an equalization plate which is essentially a perforated plate. Think of it as parallel orifices in a flow path. At higher velocities, the resistance (pressure drop) is greater. The higher velocity side of the velocity profile is restricted more than the lower velocity side, causing a shifting and equalization of the velocity profile. (See Figure 8.3.)

When designing a flow equalizer, it is important to realize that pressure drop can be significant, especially if the plate becomes plugged with debris. The best method for calculating pressure drop is to add the area of all the holes in the plate and determine an equivalent single hole orifice while calculating pressure drop accordingly (see Chapter 9). Be sure to note the effect of the recovery factor (Figure 9.6).

Make a schematic diagram of the compressor and adjacent piping. Note the length of straight runs of pipe, elbows, flow meters, valves, suction strainers, knockout drums, silencers, flow straighteners, instrumentation, etc. This will help in resolving system-related problems.

To ensure that a proper length of straight piping run is available upstream of the compressor, refer to Chapter 6, inlet piping. Table 8.2 is for orifice meters (reference ASME "Fluid Meters" for straight runs required upstream for an orifice with a 0.5 beta ratio). For minimum straight run of pipe required, see Table 8.2 and Figure 9.2 for specifics, or refer to ASME "Fluid Meters."

CHECK MECHANICAL OPERATING DATA

Check the trend of the axial position of the rotor and compare to the trend of the compressor efficiency. Mutual changes in performance and thrust position may indicate a balance piston or interstage labyrinth seal rub. Does the balance line ΔP (flow) also follow a similar pattern? What about the thrust bearing temperature? All of these items can indicate internal damage.

TABLE 8.2 Approximate Straight Run of Pipe Required Upstream of Orifice Plate

Piping Arrangement	Diameters of Straight Run
Long radius elbow	7
Two elbows—same plane	8
Two elbows—different plane	12
Partially closed valve	25
Gate valve	6.5
Check valve	11

CLEANING AXIAL AND CENTRIFUGAL COMPRESSORS

Sometimes dirt, polymer buildup, or other substances can clog the compressor internally and seriously degrade performance. Very small amounts of dirt on axial blades alter the blade profile and degrade performance. Cleaning a compressor may be all that is required to regain "like new" performance.

ORGANIC ABRASIVES

An axial or centrifugal compressor can be easily cleaned during normal operation (design speed) by using either uncooked rice or walnut shells as an abrasive agent. Depending on the process for which the compressor is used and the extent of contamination permitted to the process, the compressor can be cleaned with air going to the process or through the atmospheric blowoff valve. This must be determined by the user. Proceed as follows:

1. Locate a suitable piping or instrument connection at the inlet pipe after the filter where a funnel with a 1/2-inch ID spout or smaller can be introduced. This connection will be used to introduce the abrasive material and, therefore, would preferably be located at the top centerline of a horizontal piping run. This will allow the abrasive agent to fall into the air stream and prevent any accumulation of material at a point where it may be pulled into the machine as one large mass.

   If no suitable connection is available, one of the second stage filter elements may be removed to allow the introduction of material. When using this method, be certain that personnel entering the filter house have no loose objects or clothing, rags, pens, etc., which may be sucked into the compressor suction.

   If suction pressure is above atmospheric pressure, an ejector may be used to inject the material.
2. Position an operator where he can monitor the air flow and discharge temperature through the machine. Note the air flow and discharge temperature through the machine prior to introducing any foreign material. Vibrations should be monitored for any abnormality during the cleaning. If any vibration increases suddenly, suspend the cleaning immediately until the source of the vibration is determined and corrected.
3. Slowly begin introducing the walnut shells or uncooked rice through the funnel at a rate not to exceed *one pound per minute*. The largest dimension of the abrasive should be no greater than one-fourth inch, or one-half the size of the smallest passageway in the compressor, whichever is smaller.
4. Stop introducing the abrasive after five pounds and check the difference in air flow and discharge temperature through the machine (this assumes that speed, discharge pressure and stator vane setting have remained constant). Note the variation in flow and temperature from the flow noted in item 2.

Repeat steps 2, 3, and 4 until no further upward change is noted in air flow through the machine and a decrease in discharge temperature is observed.

Depending on the internal cleanliness of the machine, the air flow rate may change rapidly with the introduction of the abrasive (walnut shells or uncooked rice). Therefore, if the machine is on-line to the process, care must be taken that a process upset does not occur due to large changes in air flow rates.

When introducing the abrasive into the air flow stream, be certain to introduce the material in a steady flow. It is important that the abrasive not enter the air stream as a large mass, clump, or batch. This will cause damage to the blading.

Proper preparation by the persons introducing the material will avoid problems. It is suggested that a "dry run" be made prior to the actual cleaning to determine the funnel size best suited to the abrasive particle size used. This will prevent plugging of the funnel or "batch" introduction of abrasive material and help to establish a better understanding of exactly what is required.

LIQUID WASH

For cleaning of centrifugal compressors, note that the above procedure will only clean the gas passages. It will not clean between the impeller(s) and diaphragm (or backplate). Material buildup in this area can result in a rub and high vibration. In order to clean this area, a liquid wash or steam cleaning is required.

To liquid wash, fill the compressor with a suitable solvent (hot condensate) almost to the horizontal centerline, but below the seal level. A manometer can be used for this. Provide buffer to seals so liquid does not get into bearing and seal cavities.

While injecting the liquid solvent, rotate the shaft at 20 rpm or less. After 15 minutes, drain all liquid from casing. Sample the liquid drained from the casing for suspended and dissolved particles. Repeat the process until the sample shows suspended and dissolved particles are at a minimum.

For single-stage compressors, it is most effective to inject liquid solvent or steam via a port between the backplate and impeller. Liquid through spray nozzles or steam may be injected to this area on a continuous basis to minimize buildup in this area. Be cautious, however, of erosion effects.

For cracked gas compressors, coke oven gas compressors, and other applications where buildup on the diffuser surface is a problem, continuous flushing with solvent may be required. For continuous duty, a maximum solvent injection should be no more than 3% of compressor mass flow. Steam has been successfully used as a solvent for some situations such as coke oven gas. To minimize polymer buildup even water has been used. This may help by providing a "wet" surface and thereby dissuading adherence of the polymer. The results of evaporative cooling may also be a factor. Coating diaphragm surfaces with an anti-stick compound may provide further resistance to fouling.

For severe fouling, the compressor must be disassembled and cleaned by mechanical means and/or by a high-pressure spray gun utilizing solvents, caustic, steam, or just plain water. Be sure to pull the diaphragm and clean the return channels and all other flow areas. To ease future disassembly, place plastic or aluminum food wrap between diaphragms and casing. Be sure to check temperature/chemical compatibility.

Caution: The above abrasive cleaning procedures are not recommended for small, high-speed compressors. The high speed, light weight, and close clearances make this type of compressor very sensitive to foreign object ingestion. Organic abrasive cleaning of such a compressor could result in equipment failure. This type of compressor should only be cleaned after disassembly. There is no need to remove the rotor. Simply remove the inlet casing and clean the impeller and diffuser areas. Clean with a mild abrasive detergent and flush clean. Reassemble.

Always check with the equipment manufacturer first before any abrasive or liquid cleaning method is attempted.

INSPECTION OF COMPRESSOR

Once you are certain that you have good data, the unit is clean, and the axial position and thrust bearing temperatures have been checked, it will be necessary to open it up and look inside.

VISUAL INSPECTION

The following inspection points are to be completed while the top half of the casing is being lifted or immediately after and before any cleaning or further disassembly is done. No special tools or templates are required, as these are visual observations only.

Check for proper rotation of impeller and return channel vanes. Diffuser vanes should also be checked if they are present. Note that impeller vanes are usually backward-leaning. Right- and left-side diaphragms of double-flow compressors can be (and have been) interchanged (Figure 8.4).

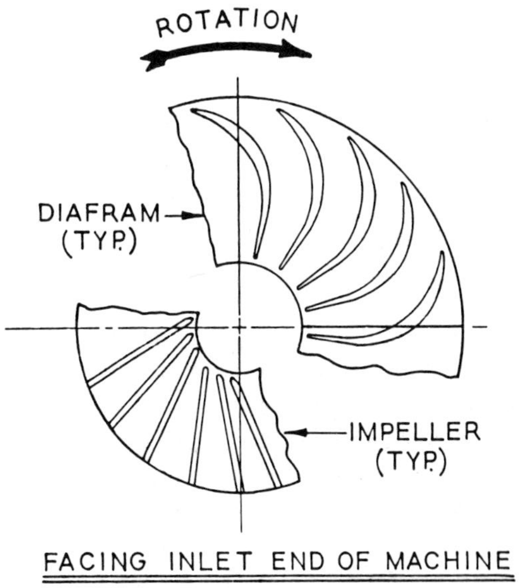

Figure 8.4. Check for proper rotation. *(Reprinted with permission of Elliott Company, Jeannette, PA.)*

Check for blockages in the flow passages. Many things have been found in compressors: T-shirts, lunch boxes, valve parts, gloves, etc.

Check to see that recirculation is not possible through the casing drain system. On barrel compressors this includes interstage leakage paths in the area between inner and outer casings.

Check the general cleanliness of the internals. Note the location and extent of rust, corrosion, and polymer buildup. If possible, take samples of the polymer, as this can help in determining the proper solvent to be used.

Note locations of erosion which may indicate that liquid or particles are being ingested. Erosion of impeller blades and diffuser passages can seriously affect performance by a change in dimension or change in surface finish.

Check for visual signs of recirculation across the diaphragm splitline. Streak marks can often be seen if there is flow across the splitline.

Some compressors, such as the Elliott horizontally split models, depend solely on the weight of the upper-half diaphragm to provide a seal at the diaphragm horizontal splitline. As the top-half casing is being lifted, check to see that the top-half diaphragms are free to fall below the horizontal splitline as far as is allowed by the anti-rotation hardware. If the diaphragms seem to be locked in place, they should be removed and cleaned so that free movement is allowed. Note the use of sealant or O-rings at the diaphragm splitlines.

Note general condition of shaft, impeller eye, and balance piston seals. Note condition of any O-rings or sealing strips that isolate the balance piston chamber area from the volute.

Check for anything that may have kept the splitline apart (dirt, tools, hardware, etc.). If sealant was used on the splitline, check for sections missing. These conditions can provide a recirculation path and degrade performance.

One area that can create problems is the balance line sleeve used in some compressors. Omission of this sleeve or O-ring during assembly can result in excessive balance piston leakage, seal oil leakage, and thrust loads (Figure 8.5).

DIMENSIONAL INSPECTION

Record shaft, impeller eye, or blade tip, and balance piston seal clearances. Top and bottom can be done by lead wire, and right- and left-side clearances by feeler gauge. Maximum and minimum clearances can be found in the operating manual.

For open wheel impellers and axial compressors, blade clearance to stationary hardware can be critical. Check

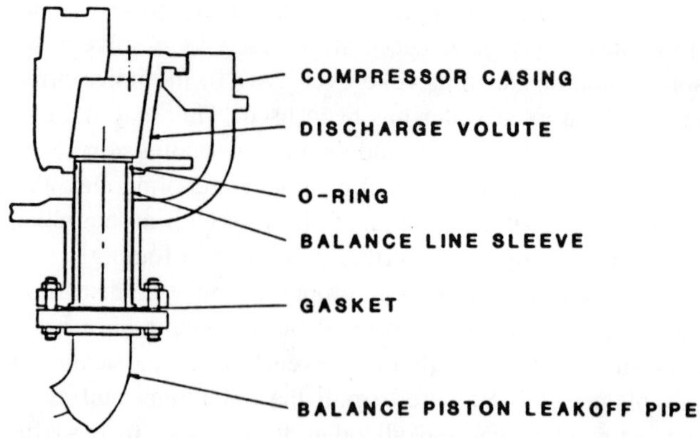

Figure 8.5. Balance piston line sleeve found in some centrifugal compressors [18]. *(Reprinted with permission of Elliott Company, Jeannette, PA.)*

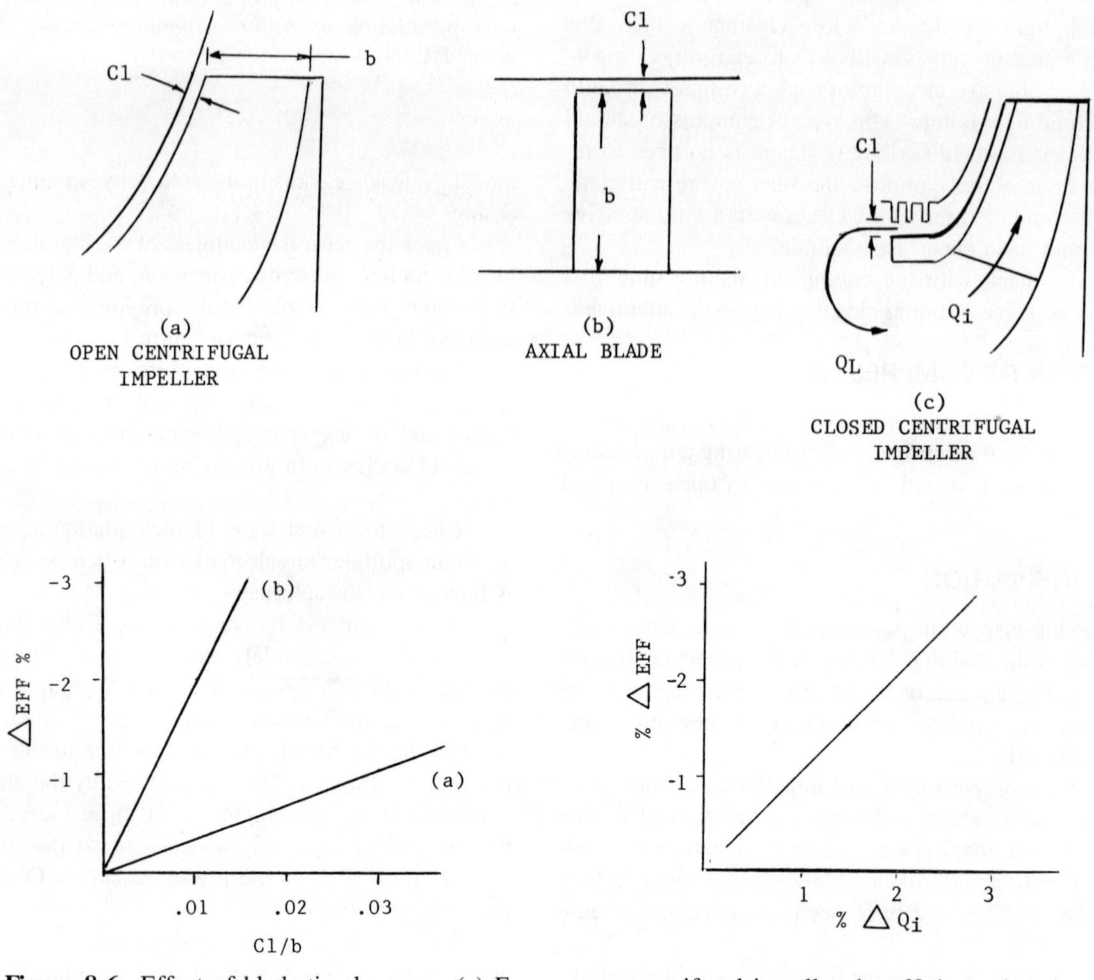

Figure 8.6. Effect of blade tip clearance. (a) For an open centrifugal impeller the efficiency loss is only one-third of a point for each percent of tip clearance ratio at the impeller outer diameter. (b) For an axial compressor blade the efficiency loss is about two percentage points for each percent of tip clearance ratio. This is also true for the throat region of an open centrifugal impeller. (c) For a closed centrifugal impeller, the efficiency loss is about one percentage point for each percent increase in Q_i (as a result of Q_L increasing and inlet flow to the compressor constant). *(Data from [27, 28].)*

clearance values according to factory specifications. The approximate effect on efficiency is shown in Figure 8.6 and the equations below [27,28].

For Axial compressors and centrifugal compressors at the impeller eye (open wheels), the compressor efficiency is reduced by 2 percentage points for each percent of tip clearance ratio.

$$\%\Delta\eta = -2 \times (\mathrm{Cl}/b)\%$$

where

$$\begin{aligned} \mathrm{Cl} &= \text{Blade tip clearance, in.} \\ b &= \text{Blade height, in.} \\ (\mathrm{Cl}/b)\% &= \frac{\mathrm{Cl}}{b} \times 100\% \end{aligned}$$

The tip clearance for centrifugal impellers is less critical. The efficiency reduction is only one-third of a point for each percent of tip clearance ratio.

$$\%\Delta\eta = -\frac{1}{3} \times \left(\frac{\mathrm{Cl}}{b}\right)\%$$

For closed centrifugal impellers with labyrinth seals at the eye, the following rule of thumb may be used:

$$\%\Delta\eta \propto -\%\Delta Q_i$$

where

$$\begin{aligned} \%\Delta Q_L &\propto \%\Delta \mathrm{Cl}_L \text{ (for small } \Delta\text{ Cl)} \\ Q_L &= \text{Labyrinth seal flow} \\ \mathrm{Cl} &= \text{Labyrinth clearance, radial} \\ Q_i &= \text{Impeller flow} \end{aligned}$$

This states that for every percent increase in impeller through-flow due to increased labyrinth leakage, the efficiency will be reduced by about 1%.

Check the axial alignment of the impellers to the diffuser walls. Tolerances and reference points (which may vary from stage to stage) are shown on the assembly drawing. The thrust bearing and housing must be in place and the endwall bolts drawn up tightly when making this check.

Pull the flow meter(s). Check for buildup or corrosion. Measure the bore. Does it check to specifications? Does the square-edged orifice have a square edge or is it rounded off from erosion? Inspect the pressure taps. Are they clean and in good condition? Inspect the pipe upstream and downstream of the flow meter. Is there any debris, dirt, or sludge buildup in the pipe? This can seriously affect performance of the flow meter.

ECONOMICS

The primary responsibility of each employee is to help keep the plant operating efficiently. That means material and labor going in one end of the plant and money coming out the other. The plant manager has a hard time justifying anything that "looks good" or "might make the equipment run better." Project requests should be in terms of money.

- How much does it cost to operate "as is" vs. the cost of operating the equipment after the proposed project is completed?
- What are projected expenses of problems that might occur if the project is not completed?
- What are costs for repairs—parts, labor, time?
- What are costs for down-time for the project?
- Are there safety concerns if the project is not completed?
- What kind of "insurance" is available? This could be in the form of spare parts, special technical help, maybe a spare machine.

Remember to check all angles and put things in terms of cash. The right dollar amount can convince the plant manager of almost anything, or it may reveal that the project is not quite as important as was originally thought.

FIELD PROBLEMS

The following is a brief summary of several actual field troubleshooting experiences.

EXPERIENCE A

Problem: During commissioning, a double-flow compressor was found to be low in head. Also, the unit surged prematurely (Figure 8.7).

Discussion: The inlet piping caused unequal flow distribution to the compressor inlet. This resulted in one section's running near surge while the other was operating near the overload region.

Resolution: The inlet piping to the compressor was modified to improve flow distribution.

EXPERIENCE B

Problem: Two duplicate single-stage air compressors were found to have a significant capacity difference during commissioning (Figure 8.8).

Discussion: Both compressors had been performance tested at the factory and were within 1.0% of each other. The suction piping for each unit was identical to the other except that they were mirror images. The axial inlet compressors had two elbows at different planes and a suction throttle valve. This piping arrangement caused flow swirl which caused prewhirl at the impeller and affected the head output.

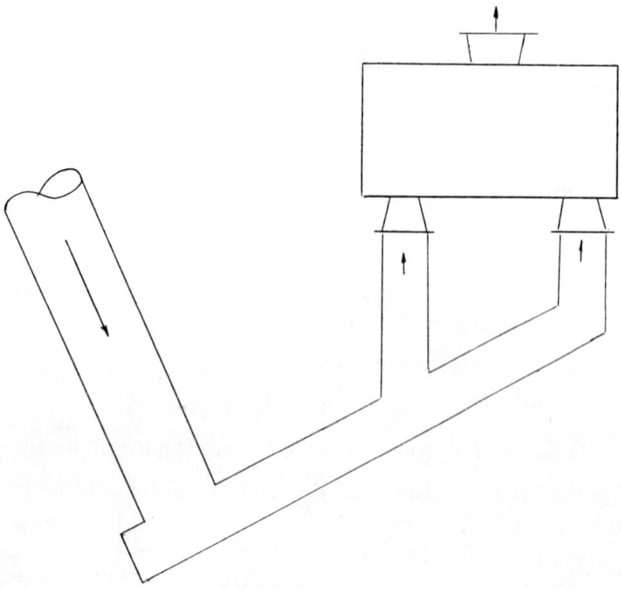

Figure 8.7

Resolution: The inlet piping was modified to include mitered elbows which minimized the problem [11].

EXPERIENCE C

Problem: During commissioning, a high-pressure multistage compressor was found to be low in head and efficiency. Near surge, control became unstable. The compressor would rumble and continue to surge even with the recycle valve open. Eventually, it would trip on high vibration (Figure 8.9).

Discussion: It was found that a vortex separator was being used upstream of the compressor to assure liquids did not enter the compressor. The residue vortex affected both the orifice and compressor performance.

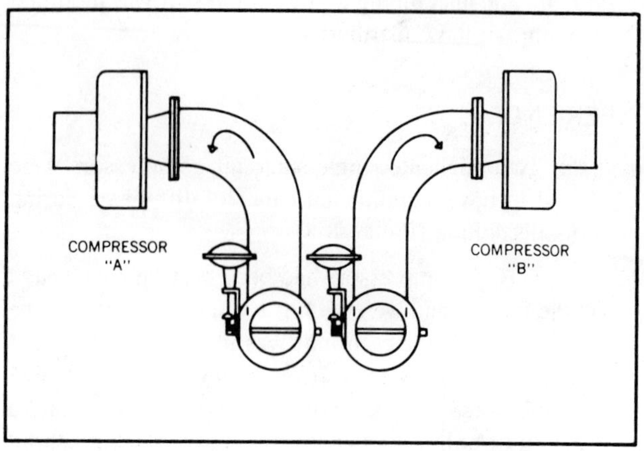

Figure 8.8

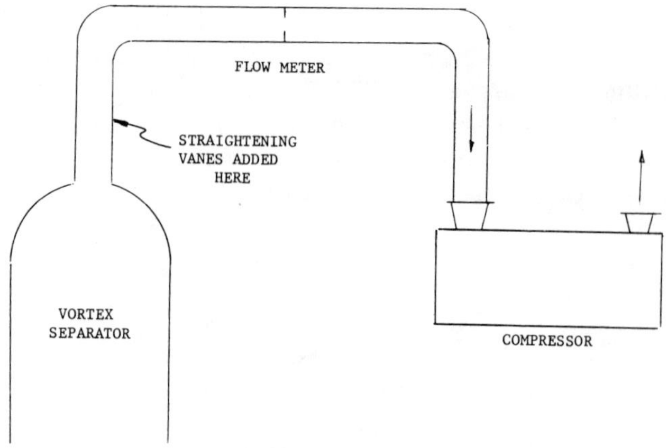

Figure 8.9

Resolution: Flow straightening vanes (egg crates) were utilized downstream of the separator to reduce the vortex.

EXPERIENCE D

Problem: Higher than normal speeds were necessary to maintain the required discharge pressure for a particular multistage compressor (Table 8.3).

Discussion: Other data were reviewed. The unit also had a high thrust bearing temperature reading. The axial position of the shaft was abnormally high, as well as the balance line ΔP. The balance piston or interstage labyrinth seals were suspected to be oversized due to a bad rub.

Resolution: The next turnaround found the impeller seals and balance piston seals damaged. Long-term solution was to install abradable seals.

EXPERIENCE E

Problem: A multi-section multistage centrifugal compressor had low head on one section (Figure 8.10).

TABLE 8.3

	Before	After
Discharge pressure (PSIG)	410	410
Discharge temperature (°F)	142	116
Axial position (Mils)	24	19
Balance line ΔP (PSID)	4.7	1.5
Speed (RPM)	11440	10770
Thrust metal temperature (°F)	240+	165

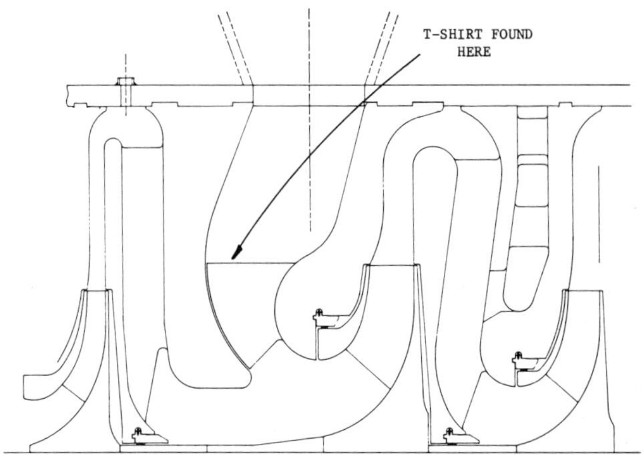

Figure 8.10

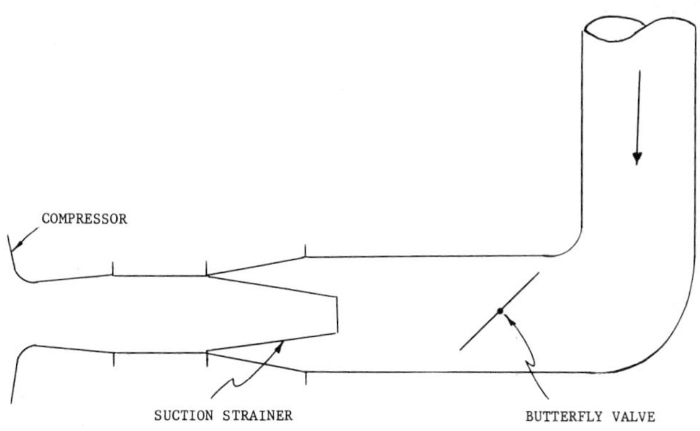

Figure 8.12

Discussion: Since the unit had just been rerated, a design or manufacturing problem was suspected. The compressor was disassembled and inspected.

Resolution: A worker's T-shirt had been found partially blocking the inlet guide vanes on the section which was low in head.

EXPERIENCE F

Problem: Upon commissioning, a multistage compressor was found to be high in capacity, causing the motor driver to overheat (Figure 8.11).

Discussion: During the factory performance test, the compressor had been found to be low in head and capacity. Adjustments were made to meet the required guarantee point.

Later it was discovered that an error had been made in the required piping upstream of the flow orifice. This piping caused flow swirl and corresponding erroneous flow measurements during the shop testing.

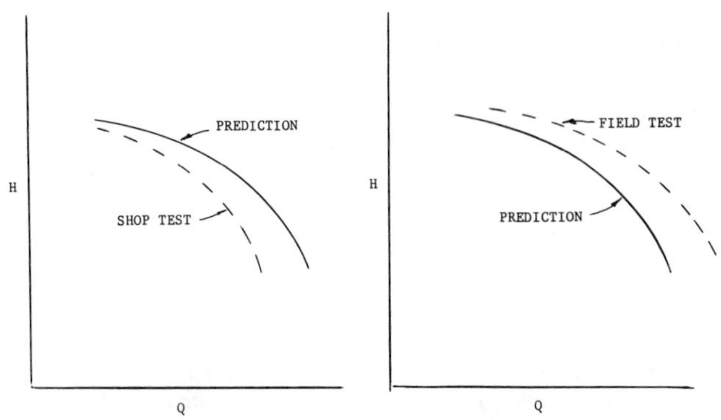

Figure 8.11

Resolution: The compressor power was reduced by changing speeds via a gear change.

EXPERIENCE G

Problem: A single-stage compressor was found to be significantly low in head (Figure 8.12).

Discussion: In addition to aerodynamic problems, the compressor had a serious mechanical problem as well. The impeller was found to be developing cracks.

The inlet piping had a strainer and a butterfly valve near the suction. Upstream of this was an elbow.

Resolution: The inlet piping was modified by replacing the suction strainer with an equalizing plate. Also, the butterfly valve was removed and put further upstream.

MAINTENANCE CHECKLIST

Check for the following items:

- Preshutdown performance
- Interstage labyrinth seal clearance
- Balance piston seal clearance
- Axial blade tip clearance
- Blade tip clearance on open centrifugal impellers
- Internal splitline leakage
- Impeller to diffuser alignment
- Cleanliness of internal parts
- Surface finish of internal parts (pitting, corrosion, buildup of foreign material, etc.)
- Proper installation of stationary guide vanes, splitter vanes, etc.
- Performance at start-up

TROUBLESHOOTING GUIDE

Symptom	Possible Cause
A. Low head	1. Low suction pressure* 2. High suction temperature* 3. Mole weight below design* 4. Internal recirculation** a. Damaged balance piston seal b. Damaged interstage seals c. Leakage across diaphragm splitline 5. Design deficiency 6. Error in test data 7. Flow swirl at suction† 8. Internal blockage or corrosion** 9. Nonuniform flow profile at suction† 10. Error in flow data 11. Impeller (or rotor blades) damaged, corroded, or dirty† 12. Diffuser (or stator vanes) damaged, corroded, or dirty** 13. Damaged or incorrect inlet or return channel vanes
B. Low capacity	1. See A above
C. High power (driver and compressor)	1. High suction pressure 2. Low suction temperature 3. High mole weight 4. Internal recirculation (see A4) 5. Internal blockage, corrosion, or dirt buildup 6. Liquid ingestion
D. Low efficiency	1. See A above
E. Poor power balance	1. High gear losses 2. Incorrect gas analysis 3. Incorrect flow data 4. Inaccurate gas properties or calculation procedure 5. Incorrect driver power
F. Premature surge	1. Increased suction pressure‡ 2. Reduced suction temperature‡ 3. Increased MW‡
G. Premature choke	1. See A above
H. Reduced operating range	1. See A above 2. See F above
I. Low power	1. Impeller blades or axial rotor blades damaged, corroded, or dirty 2. Flow swirl at suction 3. Low suction density 4. Error in test data 5. Nonuniform flow profile at suction

*See Figure 4.14.
**If work input/power is per predicted values, and head and efficiency are low, the problem is probably recirculation, blockage, or corrosion.
† If work input/power is low, the problem is centered around something that affects the ability of the impeller/rotor blades to do work—flow swirl, damaged or incorrect impeller/rotor blades, etc.
‡ See Figure 4.16.

9
FLOW METERS

Orifice and nozzle meters utilize the relationship between velocity of a flowing fluid and static pressure to measure flow. This relationship is defined by Bernoulli's equation. Simply put, as flow area is reduced (by adding a flow nozzle or orifice) the velocity must increase to maintain the same mass flow. (See Figure 9.1). Bernoulli's equation says that as the velocity is increased, the static pressure is reduced. From this, the flow equations have been developed. (See Chapter 2.)

It is most crucial that pressures and temperatures be accurately measured to assure proper flow measurement. One important factor in assuring accurate pressure measurement is that the orifice or nozzle meter be properly oriented in the piping to obtain uniform flow upstream of the orifice. According to ASME "Fluid Meters" [29], the required length of straight run of piping upstream of an orifice is shown in Figure 9.2.

SQUARE-EDGED ORIFICES

$$\dot{M} = (5.983)(K)(d^2)(\text{Fa})(Y)\sqrt{\frac{h_w}{v_1}} \quad (9.1)$$

$K = C \times E$

= Flow coefficient. If the orifice has not been individually calibrated, obtain C from Table 9.2 in this section, or from [29]. C is approximately 0.6 for high Reynolds number.

C = Discharge coefficient. Obtained from the reference tables or from manufacturer's data.

$$= K/E = K\sqrt{1-\beta^4} \quad (9.3)$$

E = Velocity of approach factor, defined as

$$\frac{1}{\sqrt{1-\beta^4}}$$

where β is the ratio of throat diameter d to pipe diameter D. (9.4)

d = Diameter of orifice in inches.

Fa = Thermal expansion factor. Obtained from Figure 9.3 or from [29]. Generally this number is very close to 1.0 for most compressor suction conditions.

Y = Net expansion factor for square-edged orifices. Ratio of flow coefficient for a gas to that for a liquid at the same value of Reynolds number. Obtained from Figure 9.4 or by equation.

ORIFICE METER EXPANSION FACTOR

For a square-edged orifice with upstream static pressure tap, use the following equation, or Figure 9.4 [29].

$$Y = 1 - \left[(0.41 + 0.35\beta^4)\left(\frac{P_2 - P_1}{P_1}\right) \div k\right] \quad (9.5)$$

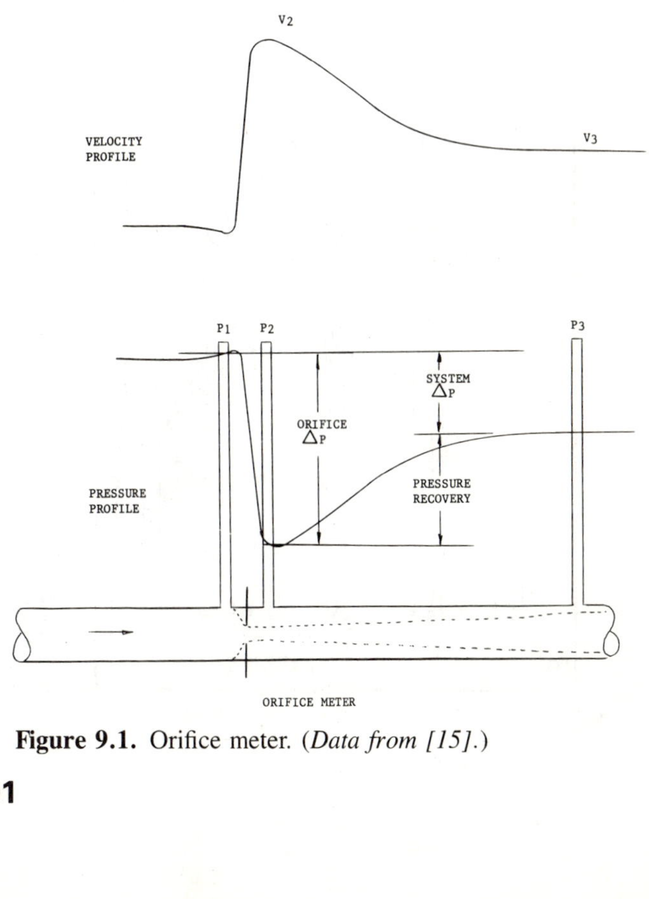

Figure 9.1. Orifice meter. (*Data from [15].*)

TABLE 9.1 Required Straight Run of Piping for Orifice

Upstream Configuration	Diameters of Straight Run Required	
	B = 0.5	B = 0.8
Long radius elbow	7	20
Two elbows—same plane	8	23
Two elbows—different plane	12	32
Partially closed valve	25	50
Full open gate valve	6.5	13.5
Check valve	11	21

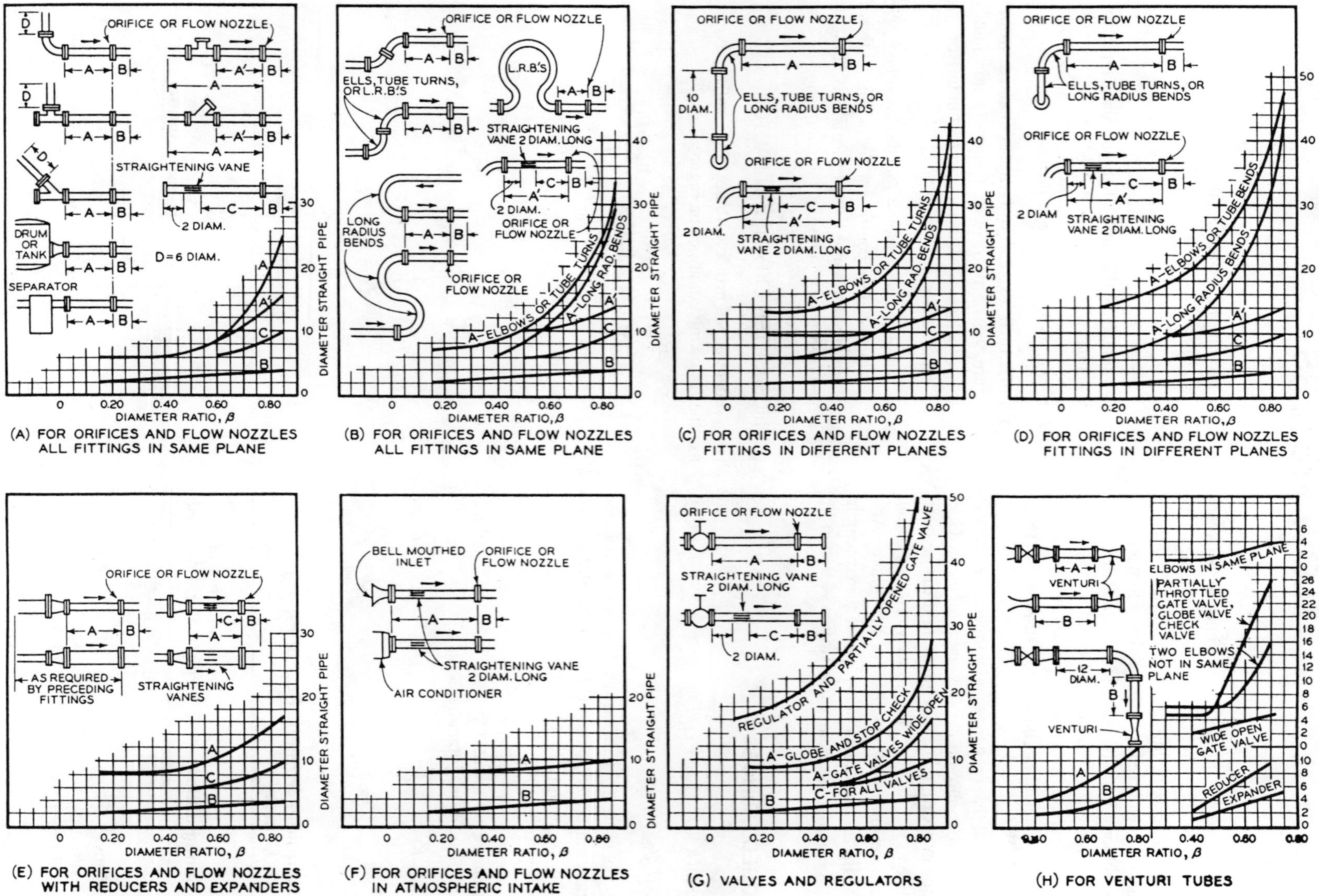

Figure 9.2. Recommended minimum lengths of pipe preceding and following orifices, flow nozzles, and venturi tubes (all control valves, including regulators, should be located on outlet side of primary element.) *(Data from [29].)*

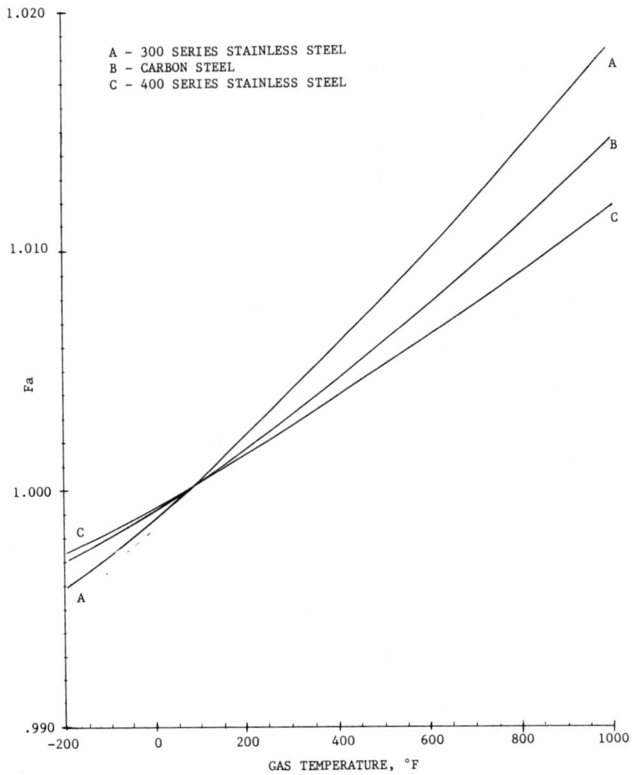

Figure 9.3. Thermal expansion factor, *Fa. (Adapted from [29].)*

ORIFICE DISCHARGE COEFFICIENT

For best results the orifice meter should be calibrated to determine the discharge coefficient. If this is not possible, Table 9.2 may be used.

FLOW NOZZLES AND VENTURI TUBES*

$$\dot{M} = (5.983)(C)(E)(d^2)(\text{Fa})(\text{Ya})\sqrt{\frac{h_w}{v_1}} \tag{9.6}$$

C = Discharge coefficient. Approximate value for flow nozzles is 0.98 for high Reynolds number. (See Table 9.3.)

E = Velocity of approach factor

$$= \frac{1}{\sqrt{1-\beta^4}}$$

d = Diameter of venturi or nozzle throat in inches

Fa = Thermal expansion factor. (See Figure 9.3.)

Ya = Expansion factor for flow nozzles or venturi tubes. (See Figure 9.5.)

*Adapted from "Fluid Meters," ASME PTC 19.5, 1971, [29].

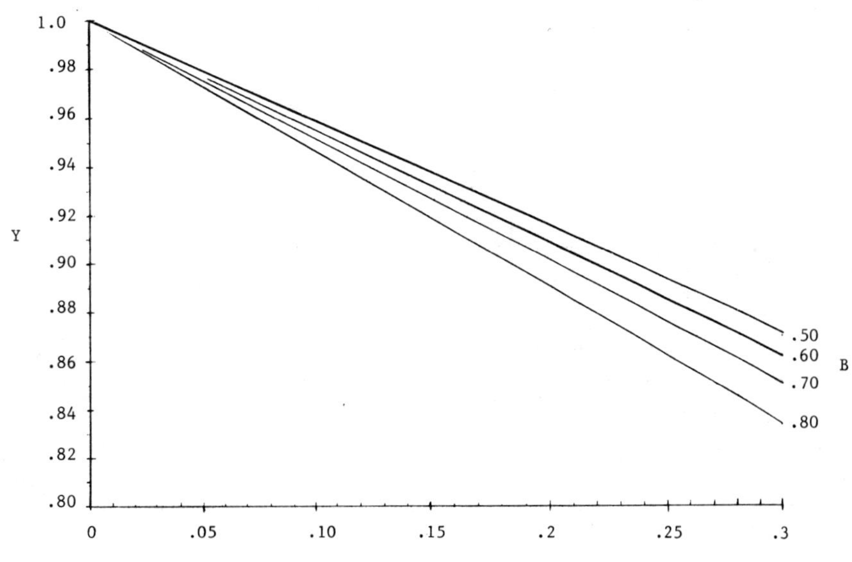

$\frac{\Delta P}{P_1} / k$

Figure 9.4. Orifice expansion factor, *Y*. Expansion factor for thin-plate square-edged orifice with flange taps, *D* and $\frac{1}{2}$ *D* taps, or vena contracta taps. Static pressure measured upstream of orifice. *(Data from [29].)*

TABLE 9.2 C, Coefficient of Discharge For Square-edged Orifices With Flange Taps*

Reynolds Number	50,000	100,000	500,000	1,000,000
β	16″ Pipe			
.5	.6197	.6110	.6039	.6031
.6	.6362	.6197	.6065	.6049
.7	.6548	.6270	.6047	.6020
.75	.6619	.6276	.6001	.5966
	2″ Pipe			
.5	.6102	.6076	.6055	.6052
.6	.6158	.6118	.6086	.6082
.7	.6183	.6124	.6077	.6071
.75	.6230	.6160	.6104	.6097

*For high Reynolds numbers, a factor of 0.6 can be assumed, and will be within ±2%. This is true for *D* and $\frac{1}{2}$ *D* taps, and vena contracta taps.
(From [29].)

FLOW NOZZLE EXPANSION FACTOR

For flow nozzles or venturi tubes use the following equation or Figure 9.5 [29].

$$\text{Ya} = \left[(r_p)^{2/k} \times \left(\frac{k}{k-1}\right) \times \left(\frac{1-(r_p)^{(k-1)/k}}{1-r_p}\right) \times \left(\frac{1-\beta^4}{1-\beta^4(r_p)^{2/k}}\right)\right]^{1/2} \tag{9.7}$$

FLOW NOZZLE DISCHARGE COEFFICIENT

For best results the flow meter should be calibrated to determine the discharge coefficient. If not, however, the following may be used.

For long-radius flow nozzles with pipe taps at 1 diameter and 1/2 diameter, use the following equation or Table 9.3 [29].

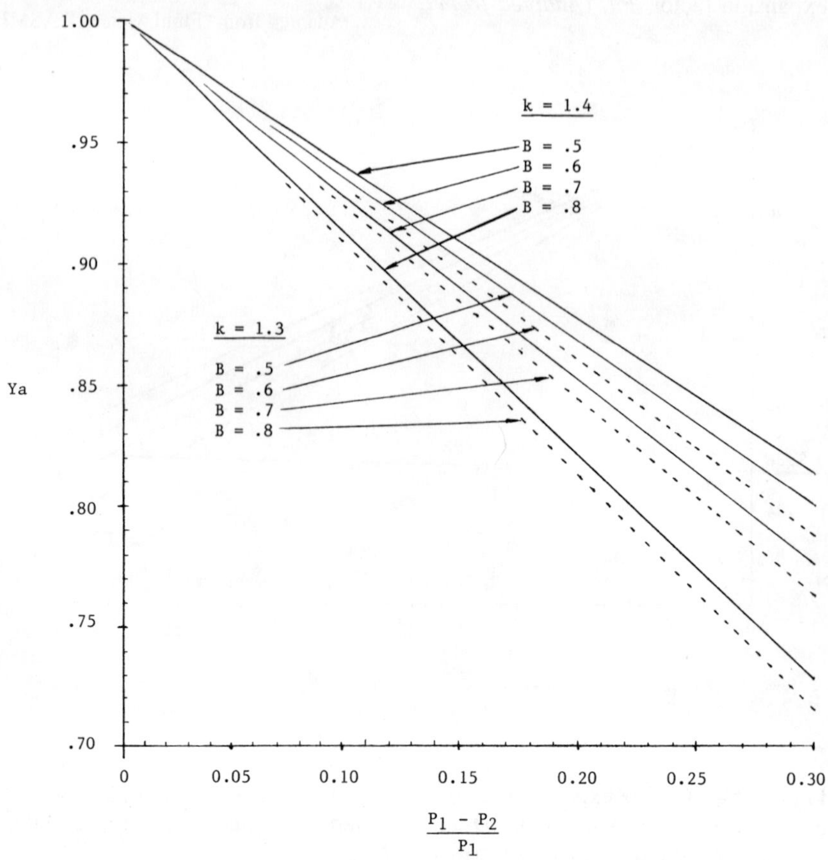

Figure 9.5. Expansion factor for flow nozzles and venturi tubes, Ya. *(Data from [29].)*

TABLE 9.3 Discharge Coefficient for Long Radius Flow Nozzles, C*

Reynolds Number	50,000	100,000	500,000	1,000,000
β	16″ Pipe			
.5	.9681	.9790	.9935	.9969
.6	.9683	.9791	.9935	.9969
.7	.9684	.9792	.9936	.9970
.75	.9685	.9792	.9936	.9970
	2″ Pipe			
.5	.9681	.9767	.9881	.9909
.6	.9682	.9767	.9882	.9909
.7	.9683	.9768	.9882	.9909
.75	.9684	.9769	.9883	.9909

*Note that a discharge coefficient of .98 can be assumed for high Reynolds number and accuracy will be within 1%. For venturi nozzles, use 0.984 for Reynolds number of 200,000 or greater.
(From [29].)

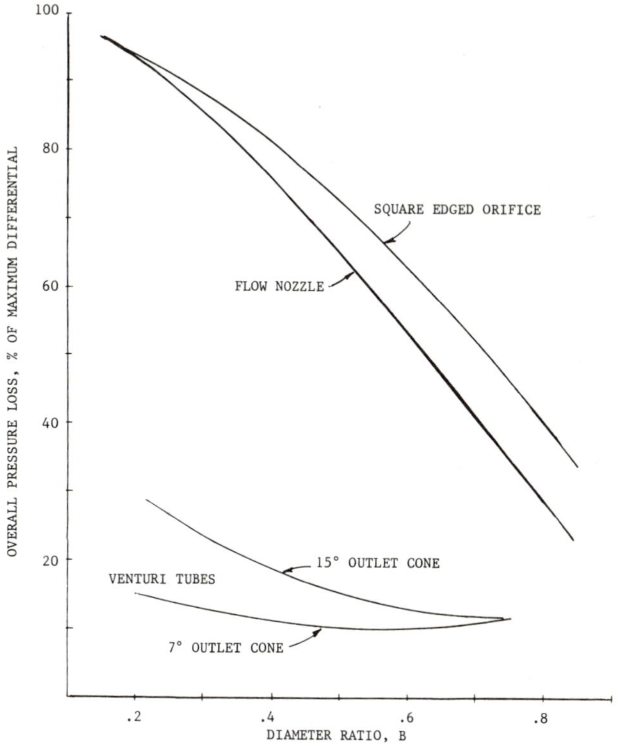

Figure 9.6. Recovery factor. *(Data from [29].)*

$$C = 0.99622 + 0.00059D - (6.36 + 0.13D - 0.24\beta^2)\frac{1}{\sqrt{\text{Re}}} \quad (9.8)$$

RECOVERY FACTOR

If the pressure were measured several pipe diameters upstream and downstream of the orifice meter or flow nozzle, the pressure difference would be found to be much less than the pressure drop measured for flow measurement. This reduced pressure drop is due to the recovery factor and is the pressure drop to be used when determining the resistance of the flow meter (Figure 9.6; see also Figure 9.1).

PITOT TUBE

The pitot tube can be a very accurate means of flow measurement if properly done. The process pipe cross-section should be divided up into several equal area sections, and a mid-point total pressure reading should be taken for each area. Preferably this is done in each quadrant to compensate for non-uniform flow (see Figure 9.7). Static pressure is measured at a normal pressure tap at the pipe wall.

These pressure measurements are then converted to velocity data.

$$P_{\text{total}} = P_{\text{static}} + P_{\text{velocity}} \quad (9.9)$$

The pitot tube measures stagnation or total pressure.

$$P_{\text{vel}} = P_{\text{total}} - P_{\text{static}}$$

$$V = \sqrt{2g_c(144)vP_{\text{vel}}} \quad (9.10)$$

After determining the fluid velocity at each equal area, the average velocity is then used to determine the flow rate.

$$Q = 60VA \quad (9.11)$$

where

Q = Flow rate, CFM
A = Pipe area, sq ft
g_c = 32.2 ft/sec^2
v = Specific volume
V = Velocity, ft/sec

Caution: The alignment of the pitot tube with the flow stream is critical. A very small angular misalignment can cause significant error. Special precautions should therefore be taken to assure precision alignment with the flow stream.

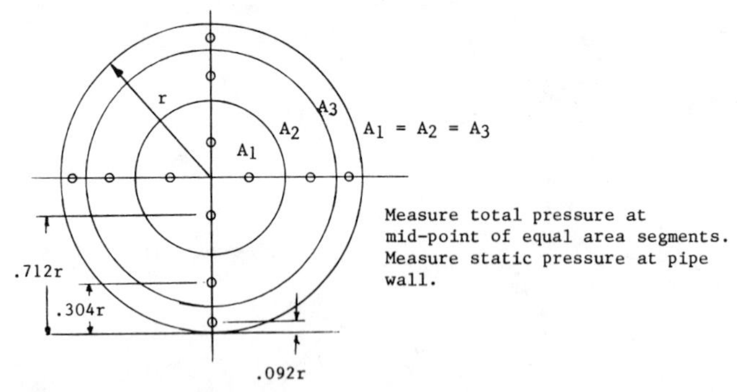

Figure 9.7. Pitot tube traverse locations.

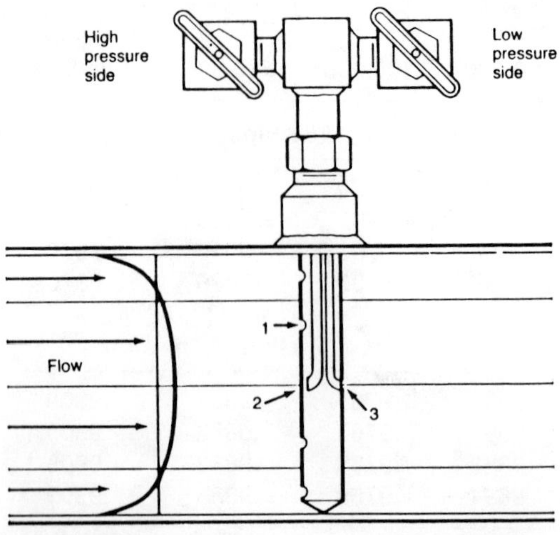

Figure 9.8. Annubar flow sensor. The annubar measuring principal is based on the pitot tube.

ANNUBAR FLOW SENSORS

A very simple device, both to install and to use, is an annubar (Figure 9.8). Using principles similar to a pitot tube, the annubar automatically averages the velocity profile by its geometry. Although not widely accepted as an accurate flow measuring device, its accuracy (or that of any flow meter) can be validated by conducting a mass flow balance on the system. This is done by measuring the flow at each inlet and discharge flange and comparing the overall mass flow rates into and out of the system. Procedures for measuring flow rates with an annubar must be obtained from the supplier.

III
REFERENCE MATERIAL

This part of the book is provided to aid in the determination of gas properties. For best results it is suggested that one of the many computer programs available for personal computers, such as the one suggested in the back of this book, be used. The data provided here, however, can produce acceptable results for many cases.

Three separate areas are covered:

APPENDIX A. GAS PROPERTIES

(Tables A.1 through A.4, and Figure A.1.) Data required for the calculation of gas properties for pure gas or gas mixtures are provided (ref. Chapters 2 and 7).

For reference, the specific volume for various gases is listed. Also, the viscosity is tabulated for selected gases at various conditions.

APPENDIX B. MOLLIER DIAGRAMS

Properties of selected gases are presented in the form of Mollier diagrams. The following gases are included:

Air

Ammonia

Argon

Carbon Dioxide

Carbon Monoxide

Chlorine

Ethane

Ethylene

Freon-11

Freon-12

Refrigerant 22

Refrigerant 500

Helium

Hydrogen

Iso-butane

Methane

Nitrogen

Oxygen

Propane

Propylene

Steam

APPENDIX C. CONVERSION FACTORS

A table of conversion factors is provided in alphabetical order.

Appendix A
GAS PROPERTIES

TABLE A.1 A. Tabulation of Gas Properties in English Units

Gas or Vapor	Hydro-carbon Reference Symbols	Chemical Formula	Molecular Mass	Specific Heat Ratio $k = c_p/c_v$ at 60°F	Critical Conditions: Absolute Pressure p_c (psia)	Critical Conditions: Absolute Temperature T_c (°R)	*$C_{p,m}$ at 50°F	*$C_{p,m}$ at 300°F
Acetylene	$C_2 =$	C_2H_2	26.05	1.24	905	557	10.22	12.21
Air		$N_2 + O_2$	28.97	1.40	547	239	6.95	7.04
Ammonia		NH_3	17.03	1.31	1636	731	8.36	9.45
Argon		A	39.94	1.66	705	272	4.97	4.97
Benzene		C_6H_6	78.11	1.12	714	1013	18.43	28.17
Iso-Butane	iC_4	C_4H_{10}	58.12	1.10	529	735	22.10	31.11
n-Butane	nC_4	C_4H_{10}	58.12	1.09	551	766	22.83	31.09
Iso-Butylene	iC_4-	C_4H_8	56.10	1.10	580	753	20.44	27.61
Butylene	nC_4-	C_4H_8	56.10	1.11	583	756	20.45	27.64
Carbon Dioxide		CO_2	44.01	1.30	1073	548	8.71	10.05
Carbon Monoxide		CO	28.01	1.40	510	242	6.96	7.03
Carbureted Water Gas (1)		–	19.48	1.35	454	235	7.60	8.33
Chlorine		Cl_2	70.91	1.36	1119	751	8.44	8.52
Coke Oven Gas (1)		–	10.71	1.35	407	197	7.69	8.44
n-Decane	nC_{10}	$C_{10}H_{22}$	142.28	1.03	320	1115	53.67	74.27
Ethane	C_2	C_2H_6	30.07	1.19	708	550	12.13	16.33
Ethyl Alcohol		C_2H_5OH	46.07	1.13	927	930	17	21
Ethyl Chloride		C_2H_4Cl	64.52	1.19	764	829	14.5	18
Ethylene	C_2-	C_2H_4	28.05	1.24	742	510	10.02	13.41
Flue Gas (1)			30.00	1.38	563	264	7.23	7.50
Helium		He	4.00	1.66	33	9	4.97	4.97
n-Heptane	nC_7	C_7H_{16}	100.20	1.05	397	973	39.52	53.31
n-Hexane	nC_6	C_6H_{14}	86.17	1.06	440	915	33.87	45.88
Hydrogen		H_2	2.02	1.41	188	60	6.86	6.98
Hydrogen Sulphide		H_2S	34.08	1.32	1306	673	8.09	8.54
Methane	C_1	CH_4	16.04	1.31	673	344	8.38	10.25
Methyl Alcohol		CH_3OH	32.04	1.20	1157	924	10.5	14.7
Methyl Chloride		CH_3Cl	50.49	1.20	968	750	11.0	12.4
Natural Gas (1)		–	18.82	1.27	675	379	8.40	10.02
Nitrogen		N_2	28.02	1.40	492	228	6.96	7.03
n-Nonane	nC_9	C_9H_{20}	128.25	1.04	345	1073	48.44	67.04
Iso-Pentane	iC_5	C_5H_{12}	72.15	1.08	483	830	27.59	38.70
n-Pentane	nC_5	C_5H_{12}	72.15	1.07	489	847	28.27	38.47
Pentylene	C_5-	C_5H_{10}	70.13	1.08	586	854	25.08	34.46
n-Octane	nC_8	C_8H_{18}	114.22	1.05	362	1025	43.3	59.90
Oxygen		O_2	32.00	1.40	730	278	6.99	7.24

TABLE A.1 A. (continued)

Gas or Vapor	Hydro-carbon Reference Symbols	Chemical Formula	Molecular Mass	Specific Heat Ratio $k = c_p/c_v$ at 60°F	Critical Conditions		*$C_{p,m}$	
					Absolute Pressure p_c (psia)	Absolute Temperature T_c (°R)	at 50°F	at 300°F
Propane	C_3	C_3H_8	44.09	1.13	617	666	16.82	23.57
Propylene	C_3-	C_3H_6	42.08	1.15	668	658	14.75	19.91
Blast Furnace Gas (1)		–	29.6	1.39	–	–	7.18	7.40
Cat Cracker Gas (1)		–	28.83	1.20	674	515	11.3	15.00
Sulphur Dioxide		SO_2	64.06	1.24	1142	775	9.14	9.79
Water Vapor		H_2O	18.02	1.33	3208	1166	7.98	8.23

(Most values taken from Natural Gas Processors Suppliers Association Engineering Data Book—1972, Ninth Edition.)
(1) Approximate values based on average composition.
* Use straight line interpolation or extrapolation to approximate $C_{p,m}$ (in Btu/mol-°R) at actual inlet T. (For greater accuracy, *average* T should be used.)
(Adapted from [30].)

TABLE A.1 B. Tabulation of Gas Properties in Metric Units

Gas or Vapor	Hydro-carbon Reference Symbols	Chemical Formula	Molecular Mass	Specific Heat Ratio $k = c_p/c_v$ at 15.5°C	Critical Conditions		*Mcp	
					Absolute Pressure p_c (bar)	Absolute Temperature T_c (K)	at 0°C	at 100°C
Acetylene	C_2-	C_2H_2	26.05	1.24	62.4	309.4	42.16	48.16
Air		$N_2 + O_2$	28.97	1.40	37.7	132.8	29.05	29.32
Ammonia		NH_3	17.03	1.31	112.8	406.1	34.65	37.93
Argon		A	39.94	1.66	48.6	151.1	20.79	20.79
Benzene		C_6H_6	78.11	1.12	49.2	562.8	74.18	103.52
Iso-Butane	iC_4	C_4H_{10}	58.12	1.10	36.5	408.3	89.75	116.89
n-Butane	nC_4	C_4H_{10}	58.12	1.09	38.0	425.6	93.03	117.92
Iso-Butylene	iC_4-	C_4H_8	56.10	1.10	40.0	418.3	83.36	104.96
Butylene	nC_4-	C_4H_8	56.10	1.11	40.2	420.0	83.40	105.06
Carbon Dioxide		CO_2	44.01	1.30	74.0	304.4	36.04	40.08
Carbon Monoxide		CO	28.01	1.40	35.2	134.4	29.10	29.31
Carbureted Water Gas (1)		–	19.48	1.35	31.3	130.6	31.58	33.78
Chlorine		Cl_2	70.91	1.36	77.2	417.2	35.29	35.53
Coke Oven Gas (1)		–	10.71	1.35	28.1	109.4	31.95	34.21
n-Decane	nC_{10}	$C_{10}H_{22}$	142.28	1.03	22.1	619.4	218.35	280.41
Ethane	C_2	C_2H_6	30.07	1.19	48.8	305.6	49.49	62.14
Ethyl Alcohol		C_2H_5OH	46.07	1.13	63.9	516.7	69.92	81.97
Ethyl Chloride		C_2H_4Cl	64.52	1.19	52.7	460.6	59.61	70.16
Ethylene	C_2-	C_2H_4	28.05	1.24	51.2	283.3	40.90	51.11
Flue Gas (1)			30.00	1.38	38.8	146.7	30.17	30.98
Helium		He	4.00	1.66	2.3	5.0	20.79	20.79
n-Heptane	nC_7	C_7H_{16}	100.20	1.05	27.4	540.6	161.20	202.74
n-Hexane	nC_6	C_6H_{14}	86.17	1.06	30.3	508.3	138.09	174.27
Hydrogen		H_2	2.02	1.41	13.0	33.3	28.67	29.03

TABLE A.1 B. (continued)

Gas or Vapor	Hydro-carbon Reference Symbols	Chemical Formula	Molecular Mass	Specific Heat Ratio $k = c_p/c_v$ at 15.5°C	Critical Conditions: Absolute Pressure p_c (bar)	Critical Conditions: Absolute Temperature T_c (*K*)	*Mcp at 0°C	*Mcp at 100°C
Hydrogen Sulphide		H_2S	34.08	1.32	90.0	373.9	33.71	35.07
Methane	C_1	CH_4	16.04	1.31	46.4	191.1	34.50	40.13
Methyl Alcohol		CH_3OH	32.04	1.20	79.8	513.3	42.67	55.32
Methyl Chloride		CH_3Cl	50.49	1.20	66.7	416.7	45.60	49.82
Natural Gas (1)		–	18.82	1.27	46.5	210.6	34.66	39.54
Nitrogen		N_2	28.02	1.40	33.9	126.7	29.10	29.31
n-Nonane	nC_9	C_9H_{20}	128.25	1.04	23.8	596.1	197.07	253.10
Iso-Pentane	iC_5	C_5H_{12}	72.15	1.08	33.3	461.1	112.09	145.56
n-Pentane	nC_5	C_5H_{12}	72.15	1.07	33.7	470.6	115.21	145.94
Pentylene	C_5-	C_5H_{10}	70.13	1.08	40.4	474.4	102.11	130.37
n-Octane	nC_8	C_8H_{18}	114.22	1.05	25.0	569.4	176.17	226.17
Oxygen		O_2	32.00	1.40	50.3	154.4	29.17	29.92
Propane	C_3	C_3H_8	44.09	1.13	42.5	370.0	68.34	88.68
Propylene	C_3-	C_3H_6	42.08	1.15	46.1	365.6	60.16	75.70
Blast Furnace Gas (1)		–	29.6	1.39	–	–	29.97	30.64
Cat Cracker Gas (1)		–	28.83	1.20	46.5	286.1	46.16	57.31
Sulphur Dioxide		SO_2	64.06	1.24	78.7	430.6	38.05	40.00
Water Vapor		H_2O	18.02	1.33	221.2	647.8	33.31	34.07

(Most values taken from Natural Gas Processors Suppliers Association Engineering Data Book—1972, Ninth Edition.)
(1) Approximate values based on average composition.
* Use straight line interpolation or extrapolation to approximate *Mcp* [in kJ/(kmol · K)] at actual inlet T. (For greater accuracy, *average* T should be used.)
(Adapted from [30].)

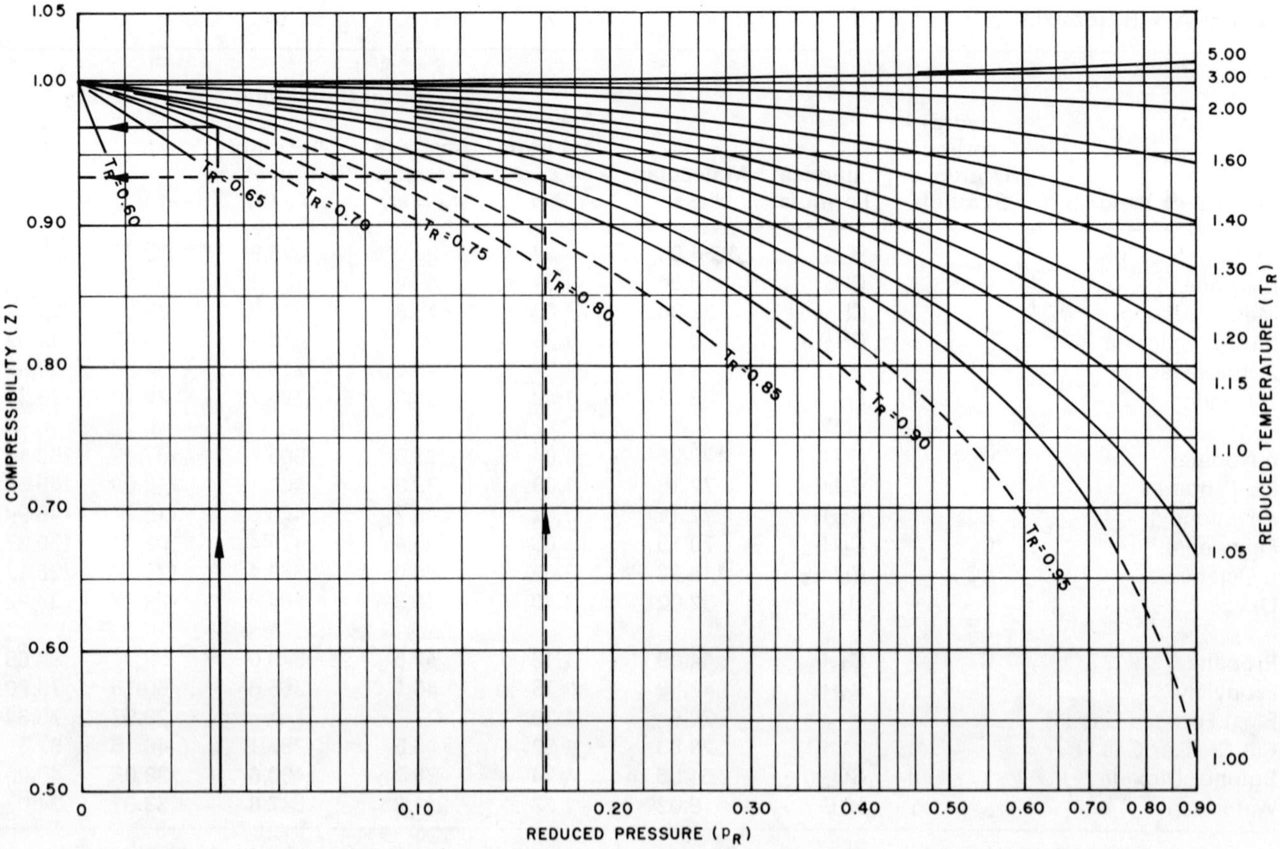

Figure A.1. Compressibility curve (Obert-Nelson). Used with data from Table A.1 and Equations (2.55) and (2.56) [30]. (*Used with permission of Elliott Company, Jeannette, PA.*)

TABLE A.2 Specific Volume of Various Gases at 14.7 PSIA, 60°F

Gas	v (ft^3/lb)
Acetylene	14.37
Air	13.09
Ammonia	22.10
Argon	9.5
Iso-butane	6.26
n-Butane	6.25
Iso-butylene	6.54
Butylene	6.54
Carbon dioxide	8.53
Carbon monoxide	13.55
Chlorine	5.25
Coke oven gas	34.10
Ethane	12.52
Ethylene	13.4
Flue gas	12.6
Helium	94.91
Hydrogen	187.8
Hydrogen sulfide	11.0
Methane	23.5
Methyl chloride	6.26
Natural gas	20.0
Nitrogen	13.53
Oxygen	11.85
Propane	8.45
Propylene	8.86
Sulfur dioxide	5.8

Source: BWR Gas Properties.

TABLE A.3 Viscosity of Various Gases*

	Viscosity, Centistokes		
Gas	0°F	200°F	400°F
Air	.0160	.0215	.0260
Ammonia	.0088	.0128	.0163
Carbon dioxide	.0130	.0181	.0225
Carbon monoxide	.0155	.0202	.0244
Chlorine	.0122	.0168	.0213
Freon-12	.0115	.0157	.0200
Helium	.0175	.0228	.0275
Hydrogen	.0080	.0101	.0121
Iso-butane	.0069	.0095	.0120
Methane	.0085	.0112	.0141
Nitrogen	.0160	.0208	.0250
Oxygen	.0181	.0240	.0290
Propane	.0074	.0101	.0128
Sulfur dioxide	.0108	.0160	.0206
Water vapor		.0107	.0136

*Approximate values for gas at or near atmospheric pressure and the temperature specified. Note that viscosity is relatively constant with pressure for pressures below 500 psi.

ν' = Kinematic Viscosity, ft^2/sec

= (Viscosity in Centistokes) $\times 1.076 \times 10^{-5}$

= (Viscosity in Centipoise) $\times 6.72 \times 10^{-4} \times v$

Source: BWR Gas Properties.

TABLE A.4 Compressor Design Operating Conditions

Compressor Operating Data		
Barometer	29.98 in. Hg. Abs.	101.35 kPa
Inlet Capacity		
Volume flow	11675 cfm	19835 m^3/h
Mass flow	5039 lb/min	38.3 kg/s
Temperature		
Inlet	122°F	50°C
Discharge	232°F	166.7°C
Maximum discharge	350°F	1.76°C
Minimum	0°F	−17.8°C
1st Iso-cool outlet	229°F	109.4°C
1st Iso-cool re-entry	110°F	43°C
Pressure		
Inlet	134.7 psia	929 kPa
Rated discharge	573.5 psia	3954 kPa
Maximum discharge	715 psia	4928 kPa
1st Iso-cool outlet	270 psia	1862 kPa
1st Iso-cool re-entry	264 psia	1822 kPa
Maximum Allowable Pressure		
Diff. across inter. diaphragm	50 psia	344.75 kPa
Guaranteed Power Input	14001 hp	10445 kW
Speed		
Guaranteed	7693 r/min	
Maximum continuous	8077 r/min	
1st Critical – book	4835 r/min	
2nd Critical – book	11885 r/min	
Gas Inlet Conditions		
Molecular mass	19.728	
Specific heat ratio (k_1)	1.247	
Compressibility (Z_1)	.981	
Gas Discharge Conditions		
Specific heat ratio (k_2)	1.228	
Compressibility (Z_2)	.967	

Appendix B

MOLLIER DIAGRAMS

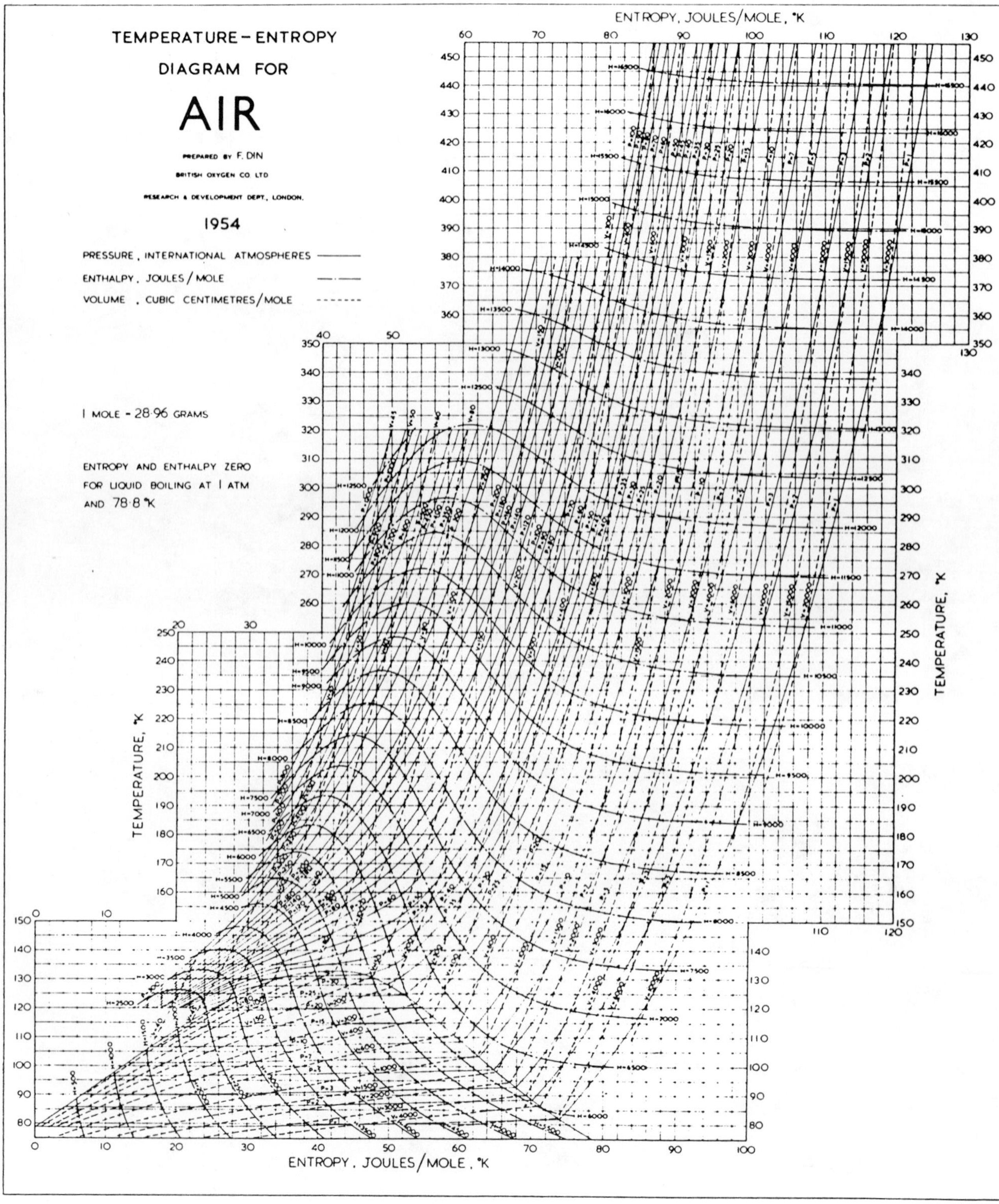

Figure B.1. Temperature-entropy diagram for air. *(From F. Din, ed.,* Thermodynamic Functions of Gases, *vol. 2, Butterworth & Co. London, 1956, with permission.)*

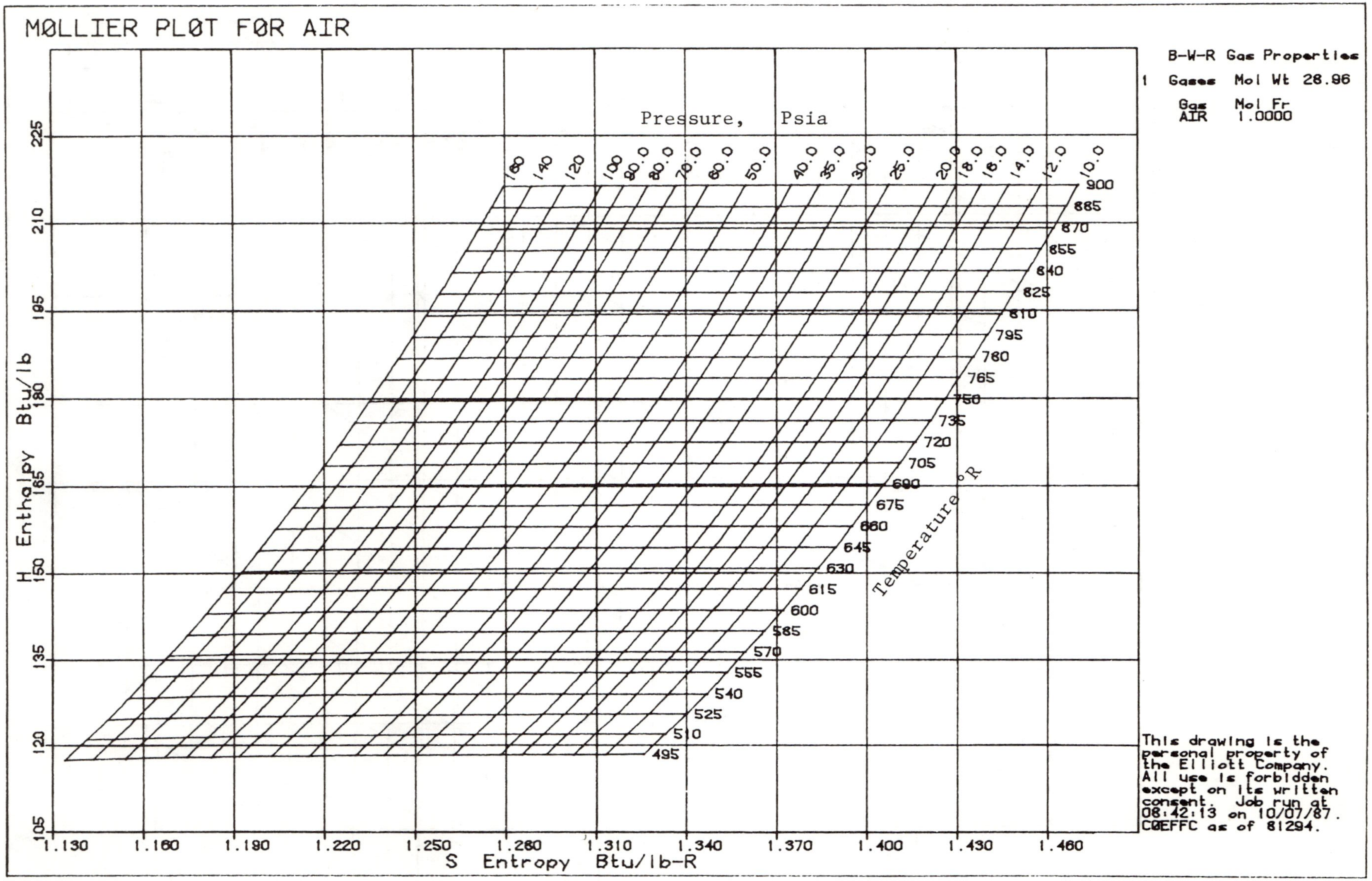

Figure B.2. Mollier plot for air. *(Used with permission of Elliott Company, Jeannette, PA.)*

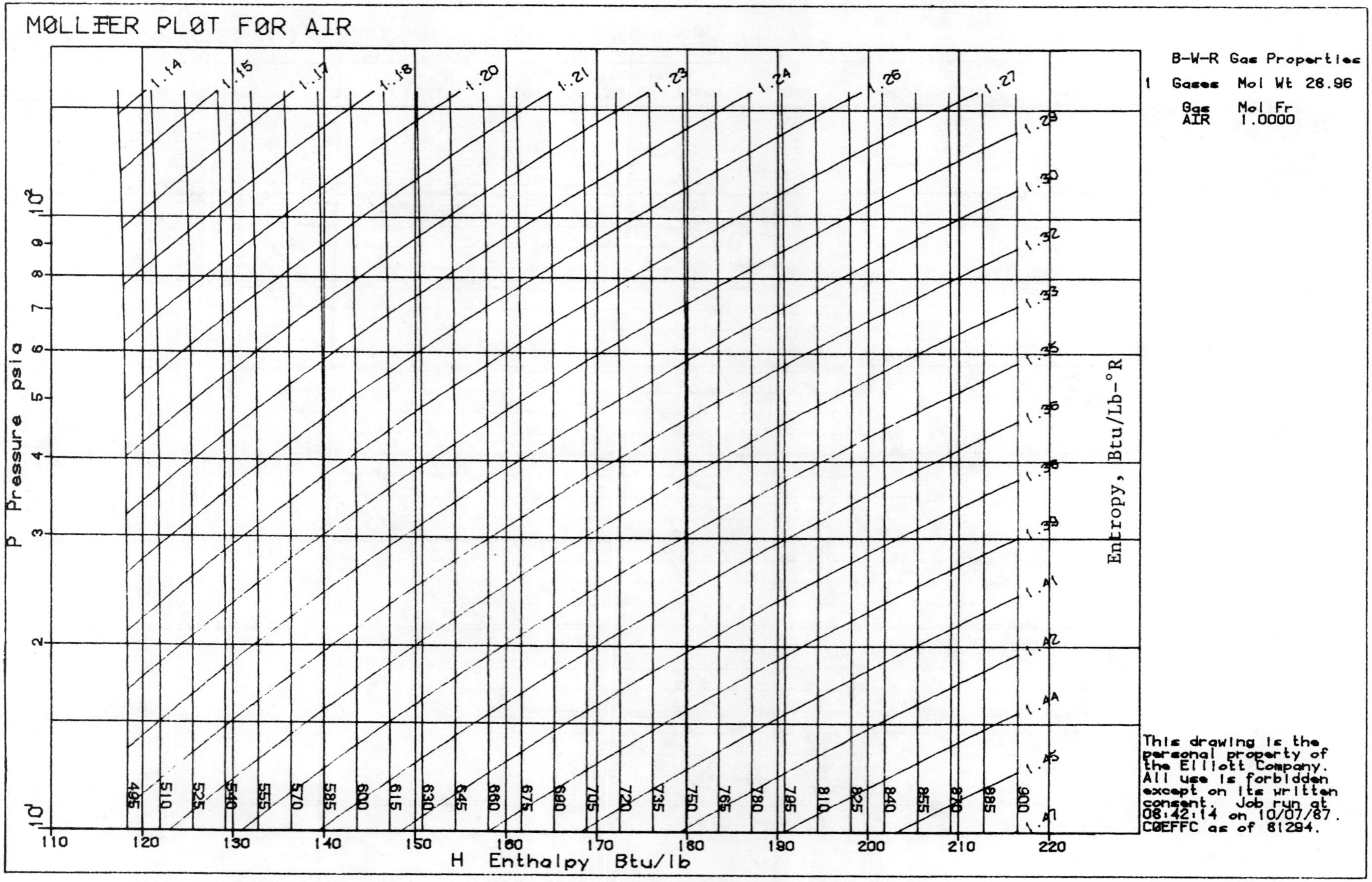

Figure B.3. Mollier plot for air. *(Used with permission of Elliott Company, Jeannette, PA.)*

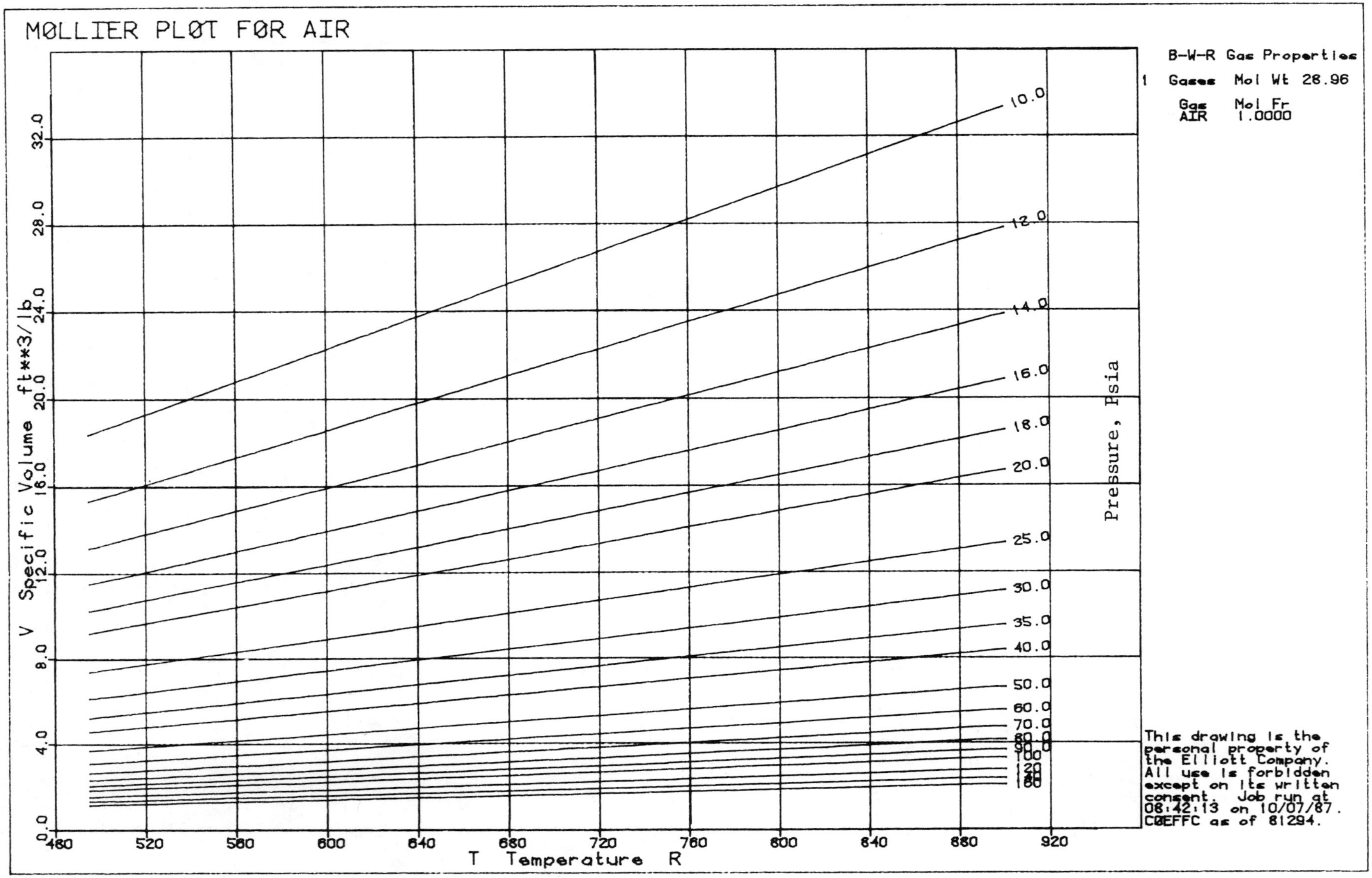

Figure B.4. Mollier plot for air. *(Used with permission of Elliott Company, Jeannette, PA.)*

Carrier® PSYCHROMETRIC CHART Normal Temperatures

Barometric Pressure
29.92 inches of Mercury

Reproduced by permission of Carrier Corporation.

ADDITIVE CORRECTIONS FOR W, h, AND v WHEN BAROMETRIC PRESSURE DIFFERS FROM STANDARD BAROMETER

Wet-Bulb Temp.	Sat. Vapor Press.	APPROXIMATE ALTITUDE IN FEET													
		−900		900		1800		2700		3700		4800		5900	
		Δp = +1		Δp = −1		Δp = −2		Δp = −3		Δp = −4		Δp = −5		Δp = −6	
t_{wb}	In. Hg	ΔW_{wb}	Δh	ΔW_{wb}	Δh	ΔW_{wb}	Δh	ΔW_{wb}	Δh	ΔW_{wb}	Δh	ΔW_{wb}	Δh	ΔW_{wb}	Δh
20	.1027	−0.5	−0.08	0.5	0.08	1.1	0.17	1.7	0.26	2.3	0.36	3.0	0.46	3.8	0.58
21	.1078	−0.5	−0.08	0.5	0.08	1.1	0.17	1.8	0.27	2.4	0.37	3.2	0.49	4.0	0.61
22	.1130	−0.5	−0.08	0.6	0.09	1.2	0.18	1.9	0.29	2.6	0.40	3.4	0.52	4.2	0.64
23	.1186	−0.6	−0.09	0.6	0.09	1.3	0.19	2.0	0.30	2.7	0.41	3.6	0.55	4.4	0.67
24	.1243	−0.6	−0.09	0.6	0.10	1.3	0.20	2.1	0.32	2.8	0.43	3.7	0.57	4.6	0.71
25	.1303	−0.6	−0.10	0.7	0.10	1.4	0.21	2.2	0.33	3.0	0.46	3.9	0.60	4.9	0.75
26	.1366	−0.7	−0.10	0.7	0.11	1.4	0.22	2.3	0.35	3.1	0.48	4.1	0.63	5.1	0.78
27	.1431	−0.7	−0.11	0.7	0.11	1.5	0.23	2.4	0.37	3.2	0.50	4.3	0.66	5.4	0.83
28	.1500	−0.7	−0.11	0.8	0.12	1.6	0.24	2.5	0.38	3.4	0.52	4.5	0.69	5.6	0.86
29	.1571	−0.8	−0.12	0.8	0.12	1.7	0.26	2.6	0.40	3.6	0.55	4.7	0.72	5.8	0.89
30	.1645	−0.8	−0.12	0.8	0.13	1.7	0.27	2.7	0.42	3.8	0.58	4.9	0.75	6.1	0.92
31	.1723	−0.8	−0.13	0.9	0.13	1.8	0.28	2.8	0.44	3.9	0.60	5.1	0.78	6.3	0.95
32	.1803	−0.9	−0.13	0.9	0.14	1.9	0.29	3.0	0.45	4.1	0.63	5.3	0.82	6.6	1.01
33	.1878	−0.9	−0.14	1.0	0.15	2.0	0.30	3.1	0.47	4.3	0.66	5.5	0.85	6.9	1.06
34	.1955	−0.9	−0.14	1.0	0.15	2.1	0.32	3.2	0.49	4.4	0.68	5.7	0.88	7.2	1.11
35	.2034	−1.0	−0.15	1.0	0.16	2.1	0.33	3.4	0.51	4.6	0.71	6.0	0.92	7.5	1.16
36	.2117	−1.0	−0.15	1.1	0.17	2.2	0.35	3.5	0.53	4.8	0.74	6.2	0.96	7.8	1.20
37	.2202	−1.0	−0.16	1.1	0.17	2.3	0.36	3.6	0.56	5.0	0.77	6.5	1.00	8.1	1.25
38	.2290	−1.1	−0.17	1.2	0.18	2.4	0.37	3.8	0.58	5.2	0.80	6.8	1.05	8.4	1.30
39	.2382	−1.1	−0.18	1.2	0.19	2.5	0.39	3.9	0.61	5.5	0.85	7.1	1.09	8.8	1.36
40	.2477	−1.2	−0.18	1.3	0.20	2.6	0.41	4.1	0.63	5.7	0.88	7.4	1.14	9.2	1.42
41	.2575	−1.2	−0.19	1.3	0.20	2.7	0.42	4.3	0.66	5.9	0.91	7.7	1.19	9.6	1.48
42	.2676	−1.3	−0.20	1.4	0.21	2.8	0.44	4.4	0.69	6.1	0.94	8.0	1.23	10.0	1.54
43	.2781	−1.3	−0.21	1.4	0.22	3.0	0.45	4.6	0.71	6.4	0.99	8.4	1.29	10.4	1.61
44	.2890	−1.4	−0.22	1.5	0.23	3.1	0.47	4.8	0.74	6.7	1.04	8.7	1.34	10.8	1.67
45	.3002	−1.4	−0.22	1.6	0.24	3.2	0.49	5.0	0.77	6.9	1.07	9.1	1.40	11.3	1.75
46	.3119	−1.5	−0.23	1.6	0.25	3.3	0.51	5.2	0.80	7.2	1.11	9.4	1.45	11.7	1.81
47	.3239	−1.6	−0.24	1.7	0.26	3.4	0.53	5.4	0.84	7.5	1.16	9.8	1.52	12.1	1.87
48	.3363	−1.6	−0.25	1.8	0.27	3.6	0.56	5.6	0.87	7.8	1.21	10.2	1.58	12.6	1.95
49	.3491	−1.7	−0.26	1.8	0.28	3.7	0.58	5.8	0.90	8.1	1.25	10.5	1.63	13.1	2.03
50	.3624	−1.7	−0.27	1.9	0.29	3.9	0.60	6.1	0.94	8.4	1.30	10.9	1.69	13.6	2.11
51	.3761	−1.8	−0.28	2.0	0.30	4.0	0.63	6.3	0.97	8.7	1.35	11.3	1.75	14.1	2.18
52	.3903	−1.9	−0.29	2.0	0.32	4.2	0.65	6.5	1.01	9.0	1.40	11.8	1.83	14.7	2.28
53	.4049	−1.9	−0.30	2.1	0.33	4.4	0.68	6.7	1.05	9.3	1.44	12.2	1.89	15.2	2.36
54	.4200	−2.0	−0.31	2.2	0.34	4.5	0.70	7.0	1.09	9.7	1.50	12.7	1.97	15.8	2.45
55	.4356	−2.1	−0.32	2.3	0.35	4.7	0.73	7.3	1.13	10.1	1.57	13.2	2.05	16.4	2.54
56	.4518	−2.2	−0.34	2.4	0.37	4.9	0.76	7.6	1.18	10.5	1.63	13.7	2.13	17.1	2.66
57	.4684	−2.3	−0.35	2.4	0.37	5.1	0.79	7.9	1.22	10.9	1.69	14.2	2.21	17.7	2.75
58	.4856	−2.3	−0.37	2.5	0.39	5.3	0.82	8.2	1.27	11.3	1.76	14.7	2.28	18.4	2.86
59	.5033	−2.4	−0.38	2.6	0.41	5.4	0.85	8.5	1.32	11.7	1.82	15.3	2.38	19.1	2.97
60	.5216	−2.5	−0.40	2.7	0.42	5.7	0.88	8.8	1.37	12.2	1.90	15.9	2.47	19.9	3.09
61	.5405	−2.6	−0.41	2.8	0.44	5.9	0.91	9.2	1.43	12.7	1.98	16.5	2.57	20.7	3.22
62	.5599	−2.7	−0.43	2.9	0.46	6.1	0.95	9.5	1.48	13.2	2.05	17.1	2.66	21.4	3.33
63	.5800	−2.8	−0.44	3.0	0.48	6.3	0.98	9.9	1.54	13.7	2.13	17.7	2.76	22.3	3.47
64	.6007	−2.9	−0.46	3.2	0.49	6.5	1.02	10.2	1.59	14.2	2.21	18.4	2.87	23.1	3.60
65	.6221	−3.1	−0.48	3.3	0.51	6.8	1.06	10.6	1.65	14.7	2.29	19.1	2.98	23.9	3.73
66	.6441	−3.2	−0.50	3.4	0.53	7.1	1.10	11.0	1.72	15.3	2.38	19.8	3.09	24.8	3.87
67	.6668	−3.3	−0.51	3.5	0.55	7.3	1.14	11.4	1.78	15.8	2.47	20.5	3.20	25.7	4.01
68	.6902	−3.4	−0.53	3.7	0.57	7.6	1.18	11.8	1.84	16.4	2.56	21.3	3.32	26.7	4.16
69	.7143	−3.5	−0.55	3.8	0.59	7.9	1.23	12.2	1.90	17.0	2.65	22.1	3.45	27.7	4.32
70	.7392	−3.7	−0.57	3.9	0.61	8.1	1.27	12.7	1.98	17.6	2.75	22.9	3.58	28.7	4.48
71	.7648	−3.8	−0.59	4.1	0.64	8.4	1.32	13.1	2.05	18.2	2.84	23.7	3.70	29.7	4.64
72	.7911	−3.9	−0.61	4.2	0.66	8.7	1.36	13.6	2.13	18.8	2.94	24.6	3.84	30.9	4.82
73	.8183	−4.1	−0.63	4.4	0.69	9.0	1.41	14.1	2.20	19.5	3.05	25.5	3.99	31.9	4.99
74	.8463	−4.2	−0.66	4.6	0.71	9.4	1.46	14.6	2.28	20.2	3.16	26.4	4.14	33.1	5.18
75	.8750	−4.4	−0.68	4.7	0.74	9.7	1.52	15.1	2.36	20.9	3.27	27.4	4.28	34.3	5.37
76	.9047	−4.5	−0.71	4.9	0.77	10.0	1.57	15.7	2.46	21.7	3.39	28.3	4.42	35.5	5.56
77	.9352	−4.7	−0.73	5.1	0.79	10.4	1.63	16.3	2.55	22.5	3.52	29.4	4.61	36.9	5.77
78	.9667	−4.9	−0.76	5.2	0.82	10.8	1.69	16.9	2.65	23.3	3.65	30.5	4.77	38.2	5.98
79	.9990	−5.0	−0.79	5.4	0.85	11.2	1.75	17.5	2.74	24.2	3.79	31.6	4.95	39.6	6.20
80	1.032	−5.2	−0.82	5.6	0.88	11.6	1.82	18.1	2.84	25.1	3.93	32.7	5.13	41.0	6.43
81	1.067	−5.4	−0.85	5.8	0.91	12.0	1.88	18.8	2.95	26.0	4.08	33.9	5.32	42.5	6.66
82	1.102	−5.6	−0.88	6.0	0.94	12.5	1.96	19.5	3.06	27.0	4.24	35.1	5.51	44.0	6.90
83	1.138	−5.8	−0.91	6.2	0.97	12.9	2.02	20.2	3.17	28.0	4.39	36.4	5.71	45.6	7.15
84	1.175	−6.0	−0.94	6.4	1.00	13.3	2.10	20.9	3.28	28.9	4.54	37.7	5.92	47.2	7.41

t_{db} = Dry-bulb temperature (F)

t_{wb} = Wet-bulb temperature (F)

p = Barometric pressure (in. of Hg)

$\Delta p'$ = Pressure difference from standard barometer (in. of Hg)

W = Moisture content of air (gr per lb of dry air)

W_{wb} = Moisture content of air saturated at wet-bulb temperature t_{wb} (gr per lb of dry air)

ΔW = Moisture content correction of air when barometric pressure differs from standard barometer (gr per lb of dry air)

ΔW_{wb} = Moisture content correction of air saturated at wet-bulb temperature when barometric pressure differs from standard barometer (gr per lb of dry air)

NOTE: To obtain ΔW reduce ΔW_{wb} by 1% where $t_{db} - t_{wb} = 24$ F

Correct proportionally, when $t_{db} - t_{wb}$ = is not 24 F

h = Enthalpy of moist air (Btu per lb of dry air)

Δh = Enthalpy correction, for saturated or unsaturated air, when barometric pressure differs from standard (Btu per lb of dry air)

= Volume of moist air (cu ft per lb of dry air)

$$v = \frac{.754\,(t_{db} + 459.7)}{p}\left[1 + \frac{W}{4360}\right]$$

Example: At a barometric pressure of 25.92 with 90 F DB and 70 F WB, determine W, h, and v. $\Delta p = -4$ and from table $\Delta W_{wb} = 17.6$. From note above,

$$\Delta W = \Delta W_{wb} - \left(\frac{20}{24} \times .01 \times 17.6\right) = 17.6 - .15 = 17.45$$

Therefore W = 78.0 (from chart) + 17.45 = 95.45 gr per lb of dry air. From table Δh = 2.75. Therefore h = saturation enthalpy from chart + deviation + 2.75 = 34.09 − .18 + 2.75 = 36.66 Btu per lb of dry air. From equation above

$$v = \frac{.754\,(90 + 459.7)}{25.92}\left[1 + \frac{95.45}{4360}\right] = 16.34 \text{ cu ft per lb of dry air}$$

ENTHALPY OF ADDED OR REJECTED MOISTURE

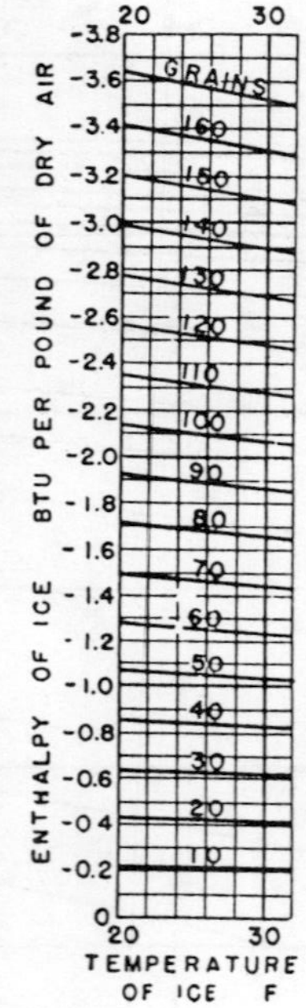

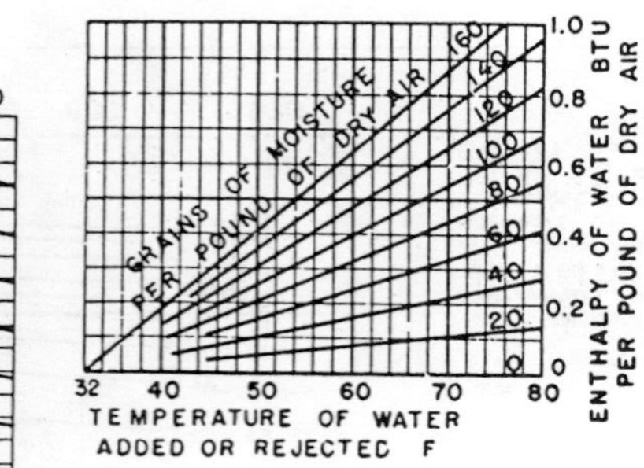

Figure B.5. Psychrometric chart for normal temperatures. *(Reproduced by permission of Carrier Corp.)*

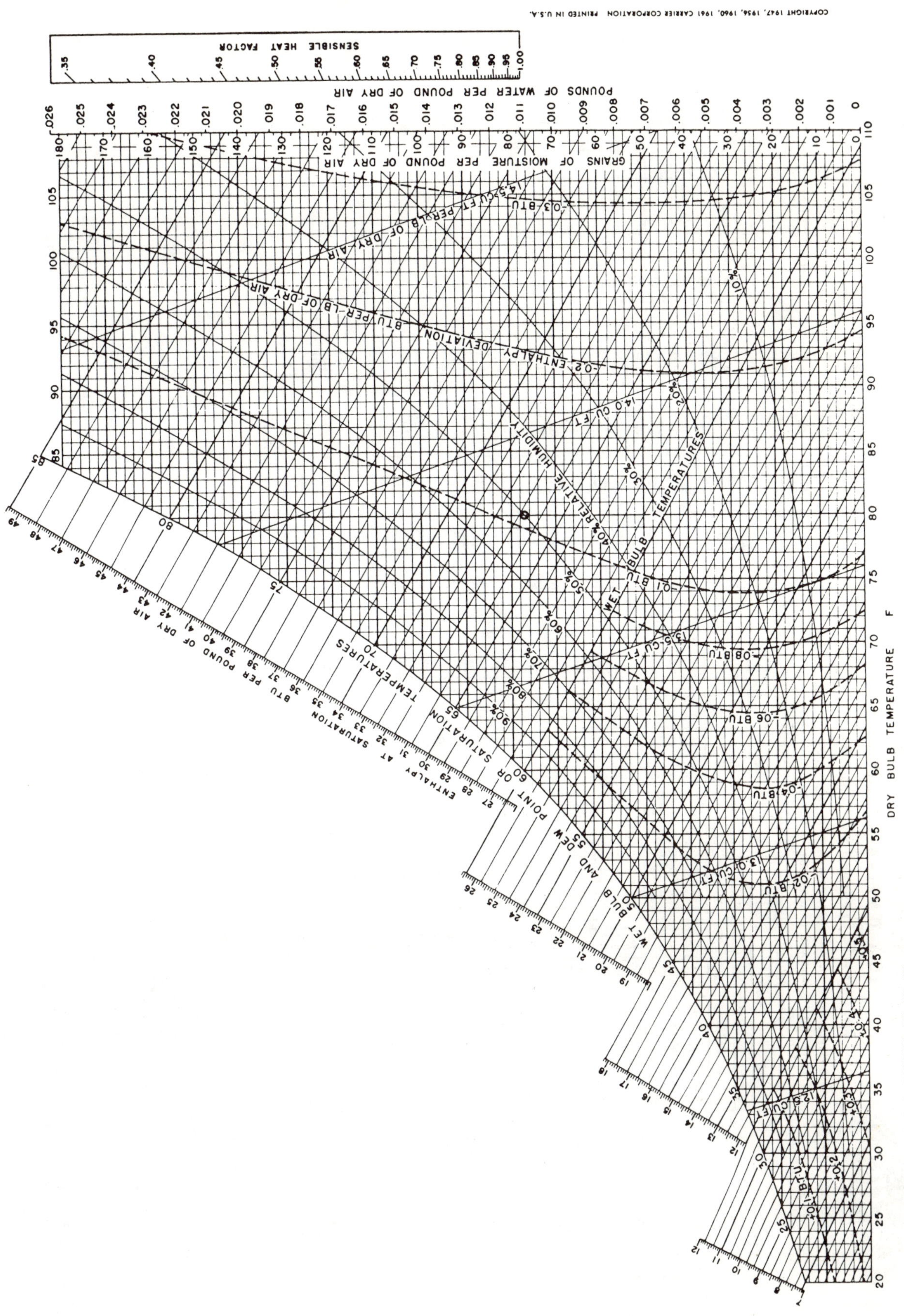

Figure B.5. (continued)

Reproduced by permission of Carrier Corporation.

Carrier®

PSYCHROMETRIC CHART

LOW TEMPERATURES

Barometric Pressure 29.92 In. Hg

ADDITIVE CORRECTIONS FOR W, h, AND v WHEN BAROMETRIC PRESSURE DIFFERS FROM STANDARD BAROMETER

Wet-Bulb Temp. t_{wb}	Sat. Vapor Press. In. Hg	APPROXIMATE ALTITUDE IN FEET													
		−900		900		1800		2700		3700		4800		5900	
		$\Delta p = +1$		$\Delta p = -1$		$\Delta p = -2$		$\Delta p = -3$		$\Delta p = -4$		$\Delta p = -5$		$\Delta p = -6$	
		ΔW_{wb}	Δh	ΔW_{wb}	Δh	ΔW_{wb}	Δh	ΔW_{wb}	Δh	ΔW_{wb}	Δh	ΔW_{wb}	Δh	ΔW_{wb}	Δh
−20	0.013	−0.06	−0.01	0.06	0.01	0.13	0.02	0.20	0.03	0.28	0.04	0.36	0.05	0.47	0.07
−18	0.014	−0.07	−0.01	0.07	0.01	0.14	0.02	0.23	0.03	0.32	0.05	0.41	0.06	0.52	0.08
−16	0.016	−0.07	−0.01	0.08	0.01	0.16	0.02	0.26	0.04	0.36	0.05	0.46	0.07	0.58	0.09
−14	0.018	−0.08	−0.01	0.09	0.01	0.18	0.03	0.29	0.04	0.40	0.06	0.52	0.08	0.65	0.10
−12	0.020	−0.09	−0.01	0.10	0.01	0.21	0.03	0.32	0.05	0.44	0.07	0.58	0.09	0.72	0.11
−10	0.022	−0.10	−0.02	0.11	0.02	0.23	0.03	0.36	0.05	0.50	0.07	0.64	0.10	0.81	0.12
−8	0.025	−0.12	−0.02	0.12	0.02	0.26	0.04	0.40	0.06	0.55	0.08	0.72	0.11	0.90	0.13
−6	0.027	−0.13	−0.02	0.14	0.02	0.29	0.04	0.44	0.07	0.62	0.09	0.80	0.12	1.00	0.15
−4	0.030	−0.14	−0.02	0.15	0.02	0.32	0.05	0.50	0.07	0.69	0.10	0.89	0.13	1.12	0.17
−2	0.034	−0.16	−0.02	0.17	0.02	0.35	0.05	0.55	0.08	0.76	0.11	0.99	0.15	1.24	0.19
0	0.038	−0.18	−0.03	0.19	0.03	0.39	0.06	0.61	0.09	0.85	0.13	1.10	0.17	1.38	0.21
2	0.042	−0.20	−0.03	0.21	0.03	0.44	0.07	0.68	0.10	0.94	0.14	1.22	0.19	1.53	0.23
4	0.046	−0.22	−0.03	0.23	0.03	0.48	0.07	0.75	0.11	1.05	0.16	1.36	0.21	1.70	0.26
6	0.051	−0.24	−0.04	0.26	0.04	0.54	0.08	0.83	0.13	1.16	0.18	1.51	0.23	1.89	0.29
8	0.057	−0.27	−0.04	0.29	0.04	0.59	0.09	0.93	0.14	1.28	0.19	1.67	0.25	2.09	0.32
10	0.063	−0.30	−0.04	0.32	0.05	0.66	0.10	1.03	0.16	1.42	0.22	1.85	0.28	2.31	0.35
12	0.069	−0.33	−0.05	0.35	0.05	0.73	0.11	1.13	0.17	1.57	0.24	2.04	0.31	2.56	0.39
14	0.077	−0.36	−0.05	0.39	0.06	0.81	0.12	1.25	0.19	1.74	0.26	2.26	0.34	2.82	0.43
16	0.085	−0.40	−0.06	0.43	0.06	0.89	0.14	1.38	0.21	1.92	0.29	2.49	0.38	3.12	0.48
18	0.093	−0.44	−0.07	0.47	0.07	0.98	0.15	1.53	0.23	2.12	0.32	2.75	0.42	3.44	0.53
20	0.103	−0.49	−0.08	0.52	0.08	1.08	0.17	1.68	0.26	2.33	0.36	3.03	0.46	3.79	0.58
22	0.113	−0.5	−0.08	0.6	0.09	1.2	0.18	1.9	0.29	2.6	0.40	3.4	0.52	4.2	0.64
24	0.124	−0.6	−0.09	0.6	0.10	1.3	0.20	2.1	0.32	2.8	0.43	3.7	0.57	4.6	0.71
26	0.137	−0.7	−0.10	0.7	0.11	1.4	0.22	2.3	0.35	3.1	0.48	4.1	0.63	5.1	0.78
28	0.150	−0.7	−0.11	0.8	0.12	1.6	0.24	2.5	0.38	3.4	0.52	4.5	0.69	5.6	0.86
30	0.165	−0.8	−0.12	0.8	0.13	1.7	0.27	2.7	0.42	3.8	0.58	4.9	0.75	6.1	0.92
32	0.180	−0.9	−0.13	0.9	0.14	1.9	0.29	3.0	0.45	4.1	0.63	5.3	0.82	6.6	1.01
34	0.197	−0.9	−0.14	1.0	0.15	2.1	0.32	3.2	0.49	4.4	0.68	5.7	0.88	7.2	1.11
36	0.212	−1.0	−0.15	1.1	0.17	2.2	0.35	3.5	0.53	4.8	0.74	6.2	0.96	7.8	1.20
38	0.229	−1.1	−0.17	1.2	0.18	2.4	0.37	3.8	0.58	5.2	0.80	6.8	1.05	8.4	1.30
40	0.248	−1.2	−0.18	1.3	0.20	2.6	0.41	4.1	0.63	5.7	0.88	7.4	1.14	9.2	1.42
42	0.268	−1.3	−0.20	1.4	0.21	2.8	0.44	4.4	0.69	6.1	0.94	8.0	1.23	10.0	1.54
44	0.289	−1.4	−0.22	1.5	0.23	3.1	0.47	4.8	0.74	6.7	1.04	8.7	1.34	10.8	1.67
46	0.312	−1.5	−0.23	1.6	0.25	3.3	0.51	5.2	0.80	7.2	1.11	9.4	1.45	11.7	1.81
48	0.336	−1.6	−0.25	1.8	0.27	3.6	0.56	5.6	0.87	7.8	1.21	10.2	1.58	12.6	1.95

t_{db} = Dry-bulb temperature (F).
t_{wb} = Wet-bulb temperature (F).
p = Barometric pressure (in. of Hg).
Δp = Pressure difference from standard barometer (in. of Hg).
W = Moisture content of air (gr per lb of dry air).
W_{wb} = Moisture content of air saturated at wet-bulb temperature t_{wb} (gr per lb of dry air).
ΔW = Moisture content correction of air when barometric pressure differs from standard barometer (gr per lb of dry air).
ΔW_{wb} = Moisture content correction of air saturated at wet-bulb temperature when barometric pressure differs from standard barometer (gr per lb of dry air).

NOTE: To obtain ΔW reduce value of ΔW_{wb} by 1% where $t - t_{wb} = 24$ F and correct proportionally when $t - t_{wb}$ is not 24 F.

h = Enthalpy of moist air (Btu per lb of dry air).
Δh = Enthalpy correction when barometer pressure differs from standard barometer, for saturated or unsaturated air (Btu per lb of dry air).
v = Volume of moist air (cu ft per lb of dry air).

$$v = \frac{.754\,(t_{db} + 459.7)}{p}\left[1 + \frac{W}{4360}\right]$$

PROPERTIES OF SATURATED AIR
−20 to −100 F

Temp. F	Vapor Press. In. Hg	Moisture Content Grains*	Volume cu ft*	Enthalpy Btu*
−20	.0126	1.83	11.08	−4.52
−22	.0112	1.63	11.03	−5.03
−24	.0100	1.45	10.99	−5.54
−26	.0089	1.29	10.93	−6.04
−28	.0079	1.15	10.88	−6.55
−30	.0070	1.02	10.83	−7.05
−32	.0062	0.90	10.78	−7.54
−34	.0055	0.80	10.73	−8.04
−36	.0049	0.71	10.68	−8.53
−38	.0043	0.62	10.63	−9.03
−40	.0038	0.55	10.58	−9.52
−42	.0033	0.49	10.53	−10.01
−44	.0029	0.43	10.48	−10.50
−46	.0026	0.38	10.43	−10.98
−48	.0023	0.33	10.38	−11.47
−50	.0020	0.29	10.33	−11.96
−52	.0017	0.25	10.28	−12.44
−54	.0015	0.22	10.23	−12.93
−56	.0013	0.19	10.18	−13.41
−58	.0012	0.17	10.13	−13.89
−60	.0010	0.15	10.08	−14.38
−62	.0009	0.13	10.03	−14.86
−64	.0008	0.11	9.98	−15.34
−66	.0007	0.10	9.93	−15.83
−68	.0006	0.08	9.87	−16.31
−70	.0005	0.07	9.82	−16.79
−72	.0004	0.06	9.77	−17.27
−74	.0004	0.05	9.72	−17.75
−76	.0003	0.05	9.67	−18.23
−78	.0003	0.04	9.62	−18.71
−80	.0002	0.03	9.57	−19.19
−82	.0002	0.03	9.52	−19.68
−84	.0002	0.03	9.47	−20.16
−86	.0001	0.02	9.42	−20.64
−88	.0001	0.02	9.37	−21.12
−90	.0001	0.02	9.32	−21.60
−92	.0001	0.01	9.27	−22.80
−94	.0001	0.01	9.22	−22.56
−96	.0001	0.01	9.17	−23.04
−98	.0001	0.01	9.12	−23.52
−100	.0000	0.01	9.07	−24.00

*Per lb of dry air

Example: At a barometric pressure of 25.92 with 30 F DB and 28 F WB, determine W, h, and v. $\Delta p = -4$ and from table $\Delta W_{wb} = 3.4$. From note above,

$$\Delta W = \Delta W_{wb} - \left(\frac{2}{24} \times .01 \times 3.4\right) = 3.4 - .003 = 3.4.$$

Therefore $W = 19.2$ (from chart) $+ 3.4 = 22.6$ gr per lb of dry air. From table $\Delta h = 0.52$. Therefore h = saturation enthalpy from chart + deviation + 0.52 = 10.1 + .06 + 0.52 = 10.68 Btu per lb of dry air. From equation above

$$v = \frac{.754\,(30 + 459.7)}{25.92}\left[1 + \frac{22.6}{4360}\right] = 14.32 \text{ cu ft per lb}$$

of dry air.

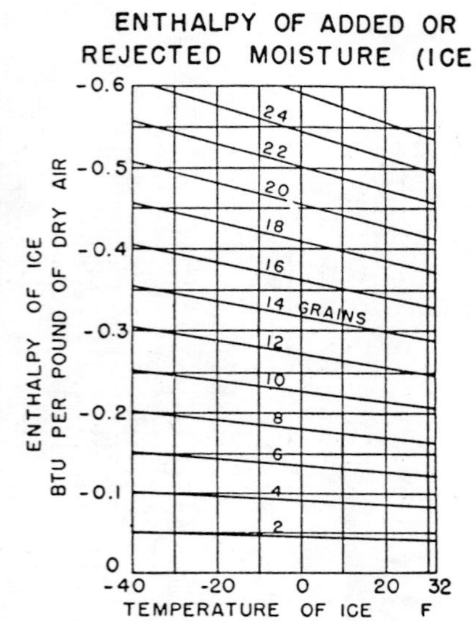

Figure B.6. Psychrometric chart for low temperatures. *(Reproduced by permission of Carrier Corp.)*

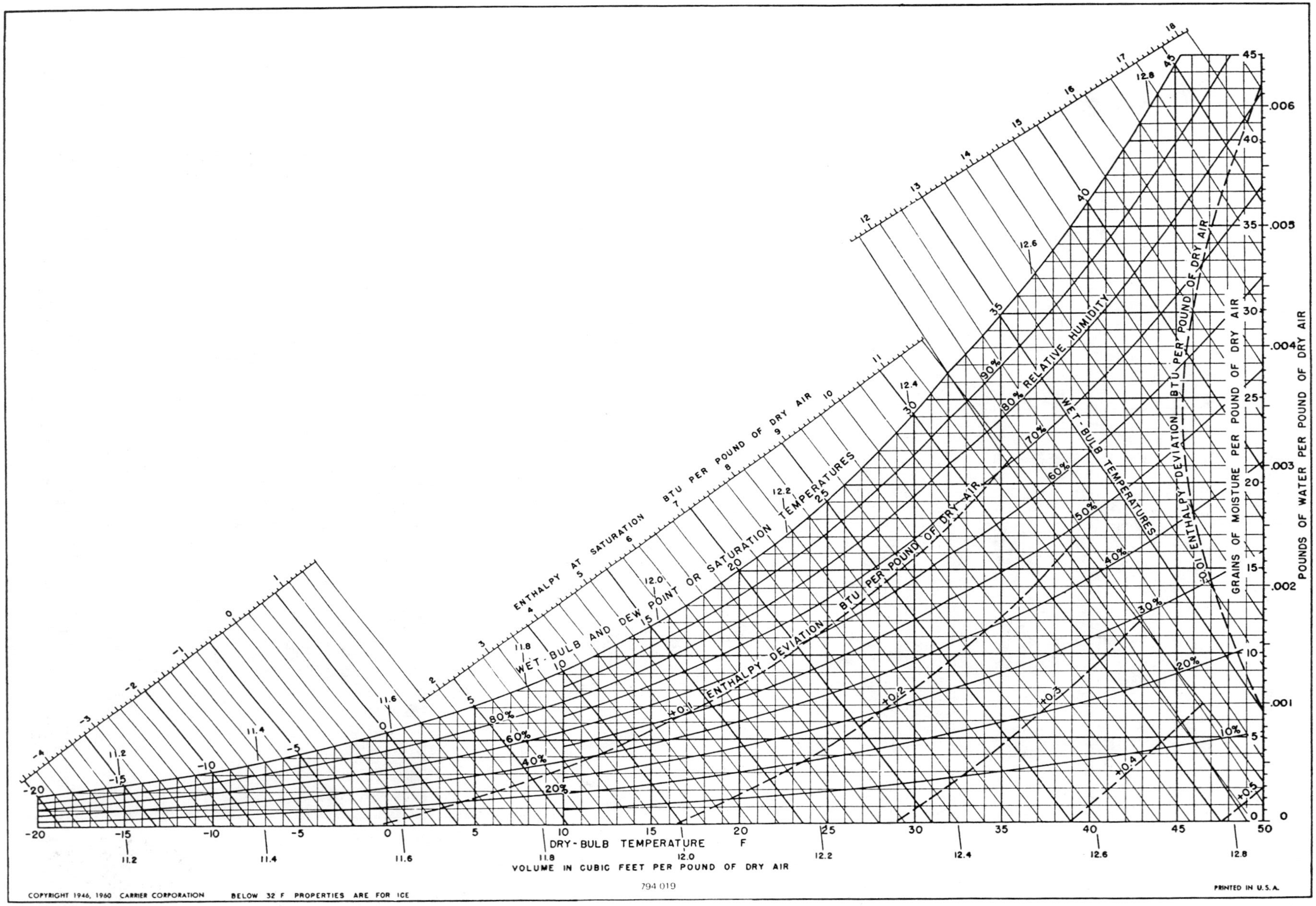

Figure B.6. (continued)

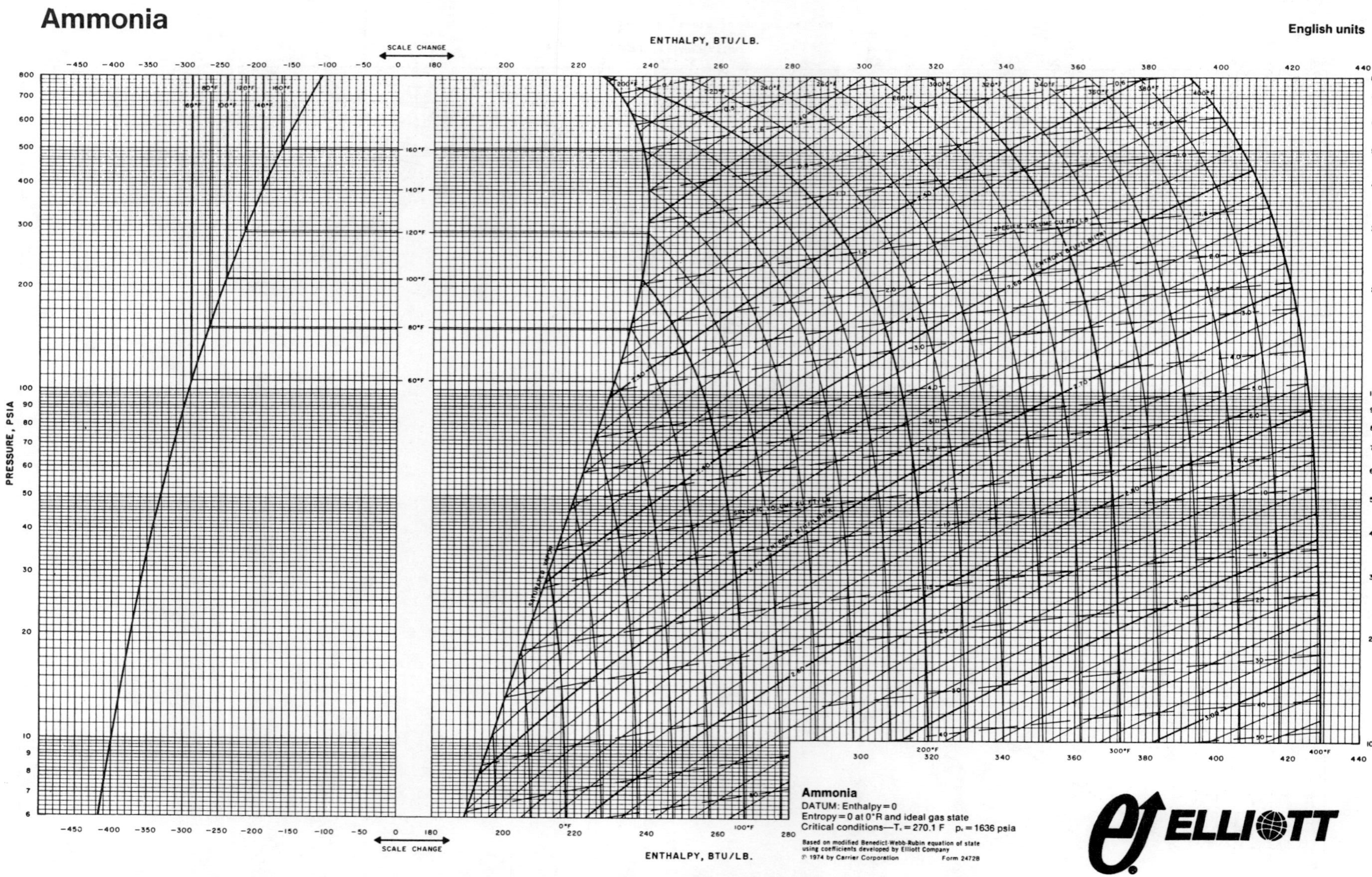

Figure B.7. Mollier plot for ammonia. *(Used with permission of Elliott Company, Jeannette, PA.)*

Ammonia

Metric units

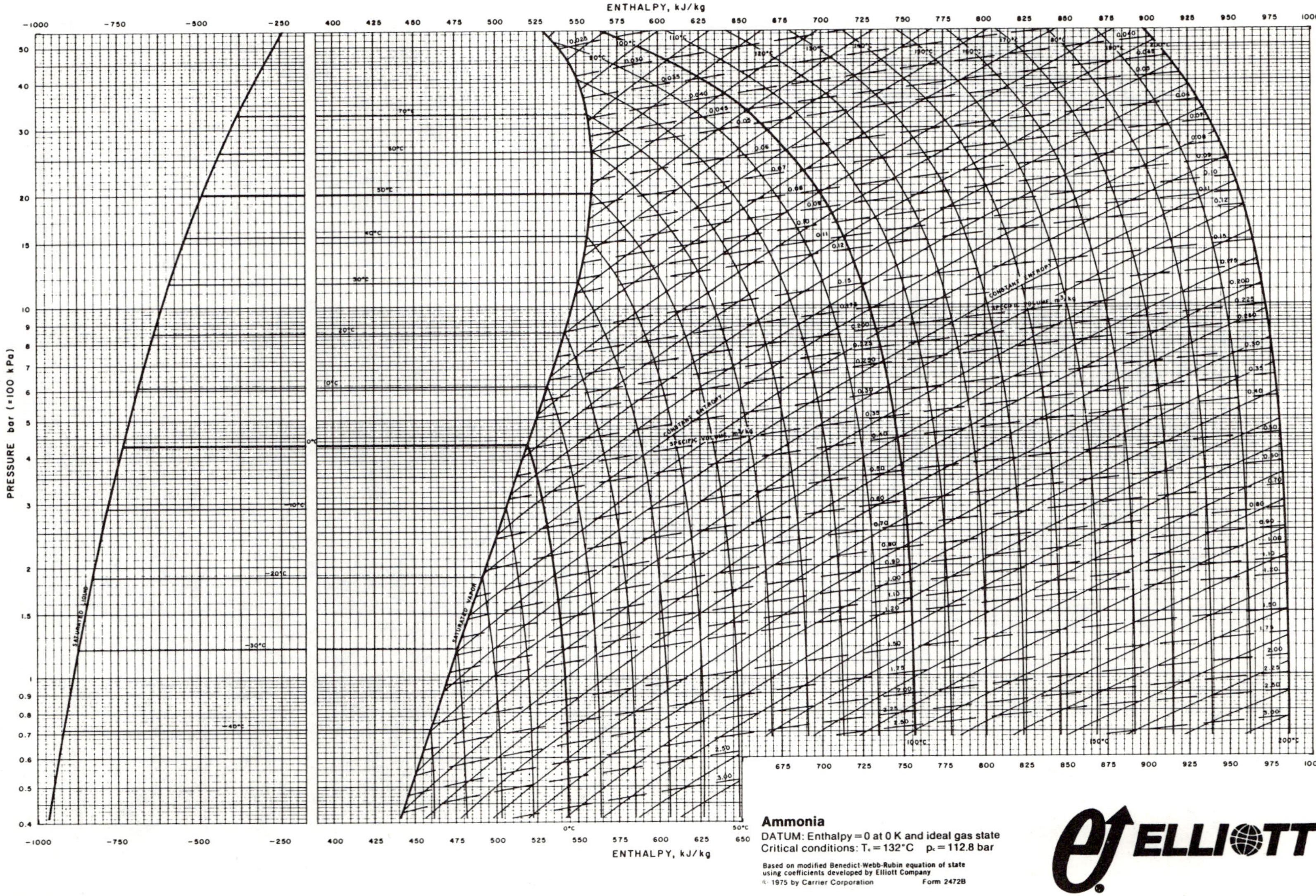

Figure B.8. Mollier plot for ammonia. *(Used with permission of Elliott Company, Jeannette, PA.)*

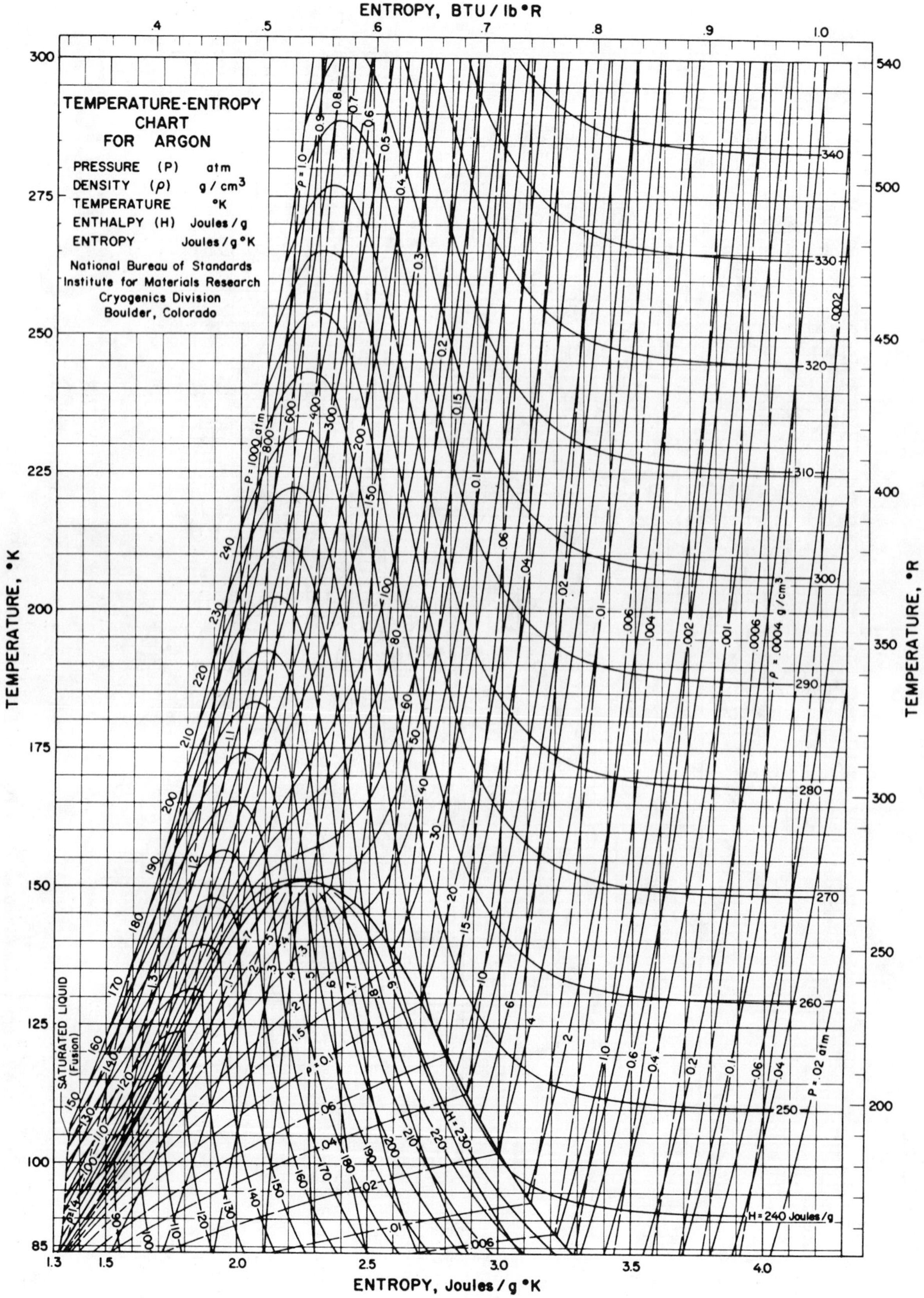

Figure B.9. Temperature-entropy diagram for argon. *(From National Bureau of Standards, Boulder, CO.)*

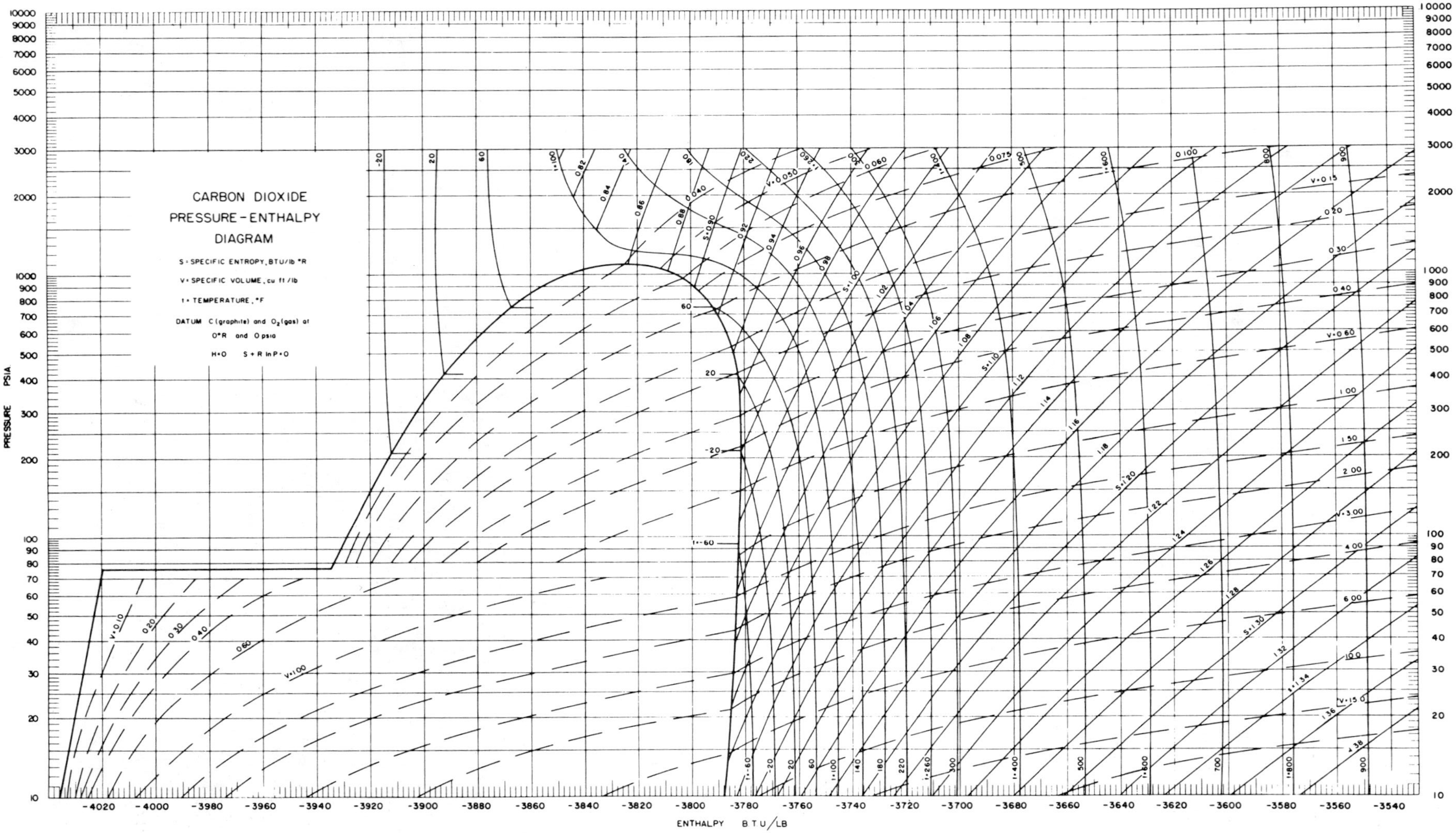

Figure B.10. *(From [31],* Thermo Properties of Non-Hydrocarbons, *by L.N. Canjar, E.K. Pollock, T.W. Cadman, W.E. Lee, and F.S. Manning. Copyright © 1966 by Gulf Publishing Company, Houston, TX. Used with permission. All rights reserved.)*

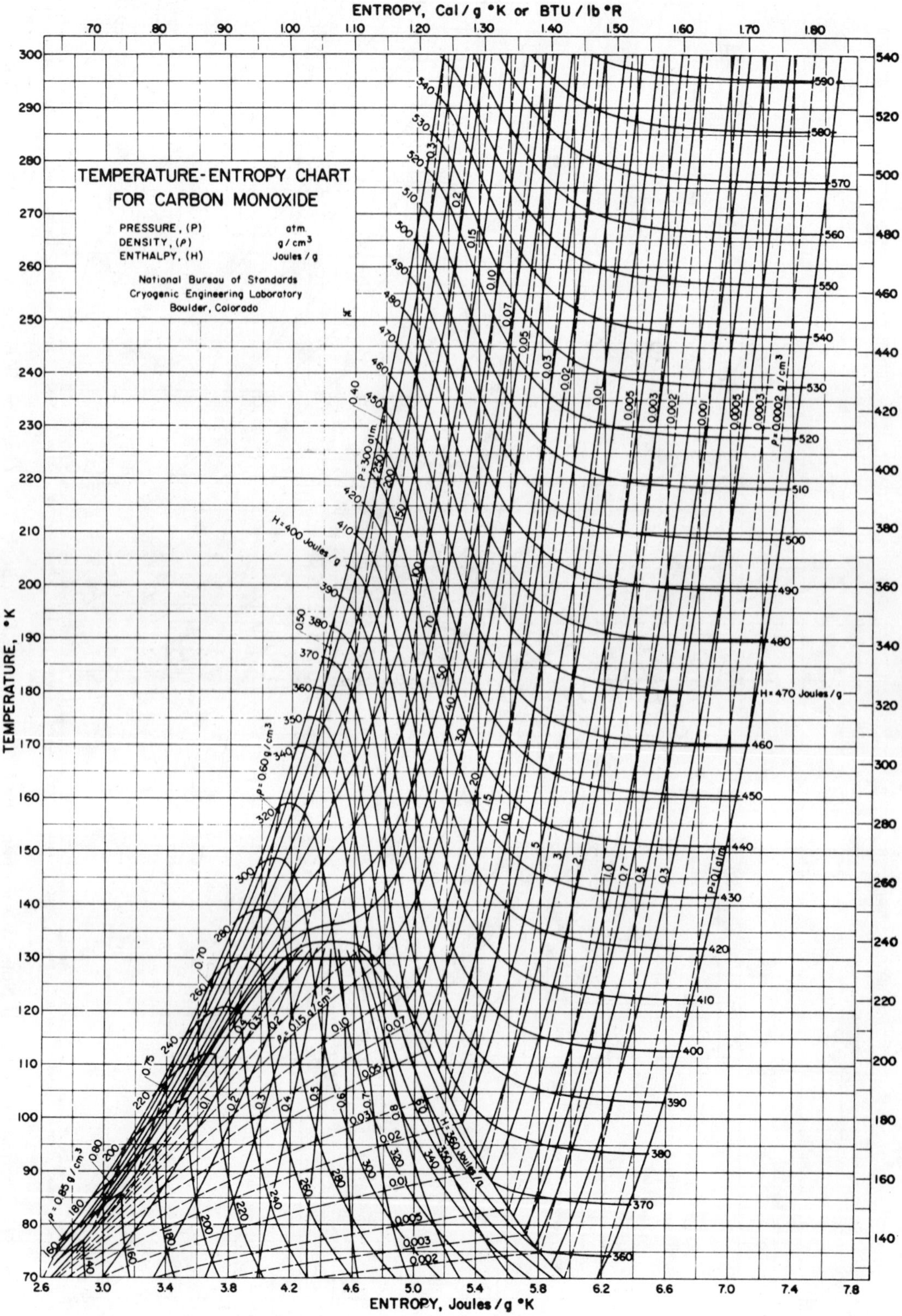

Figure B.11. Temperature-entropy chart for carbon monoxide. *(From the National Bureau of Standards, Boulder, CO.)*

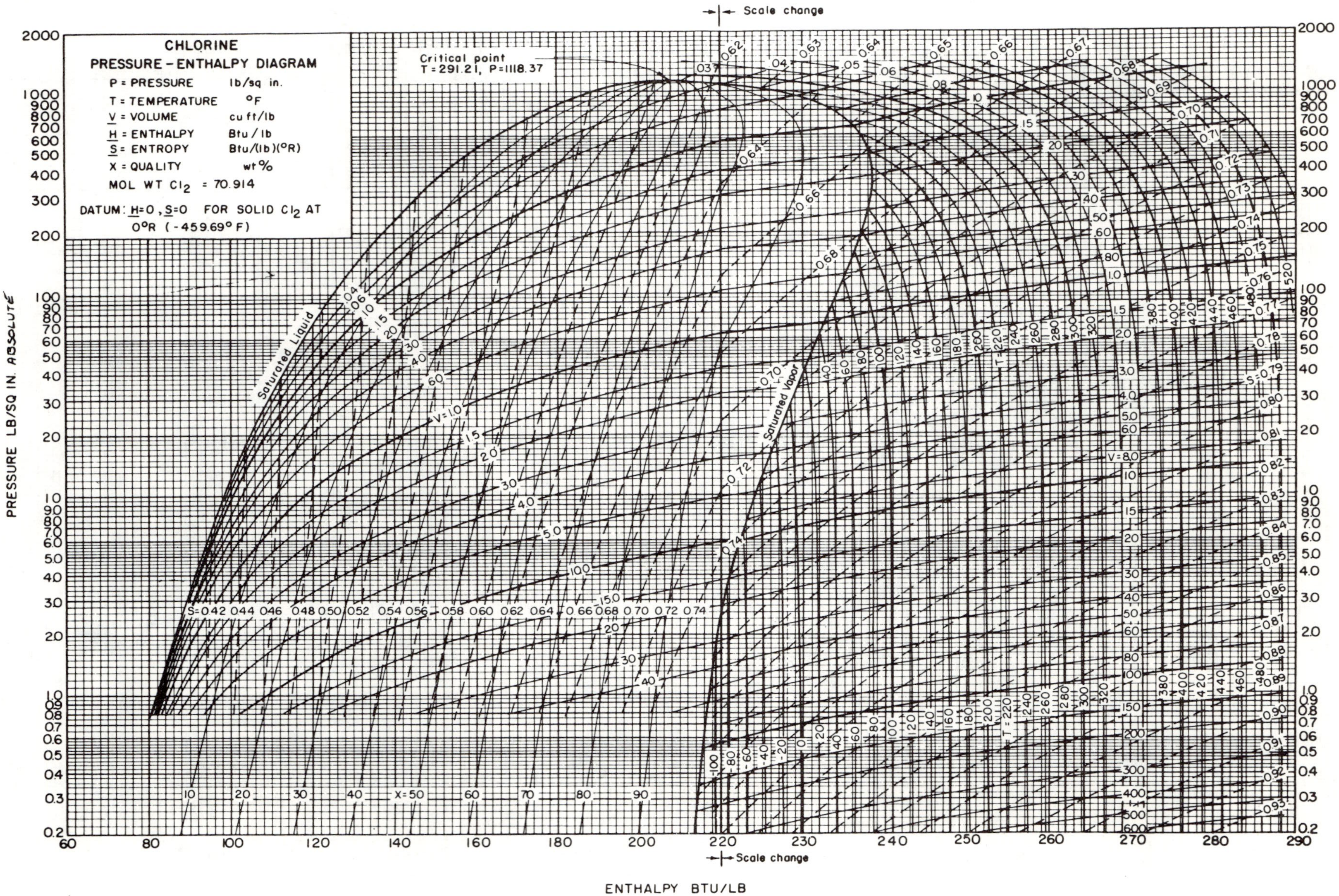

Figure B.12. Pressure enthalpy diagram for chlorine.

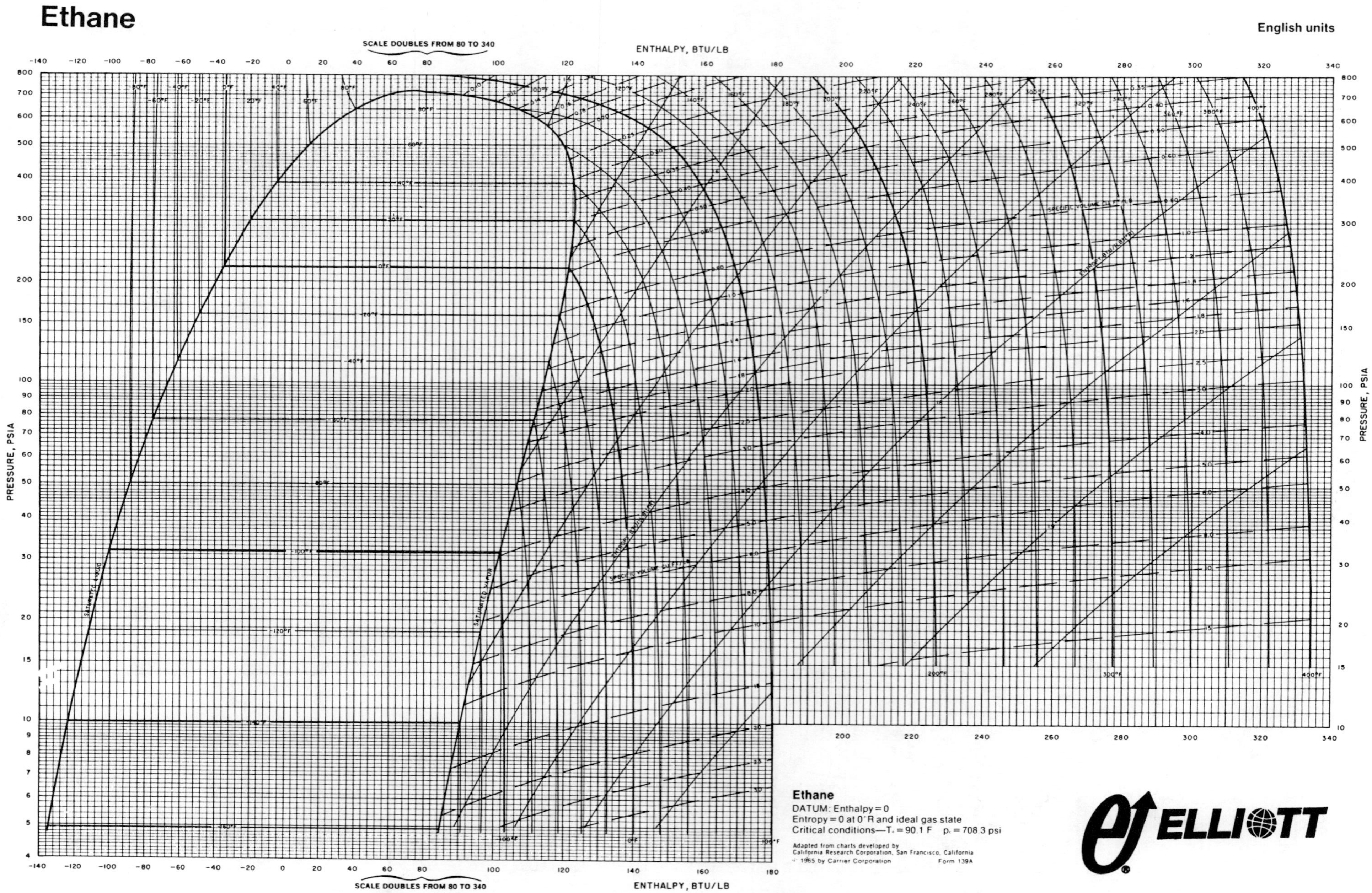

Figure B.13. Mollier diagram for ethane. *(Used with permission of Elliott Company, Jeannette, PA.)*

Ethane

Metric units

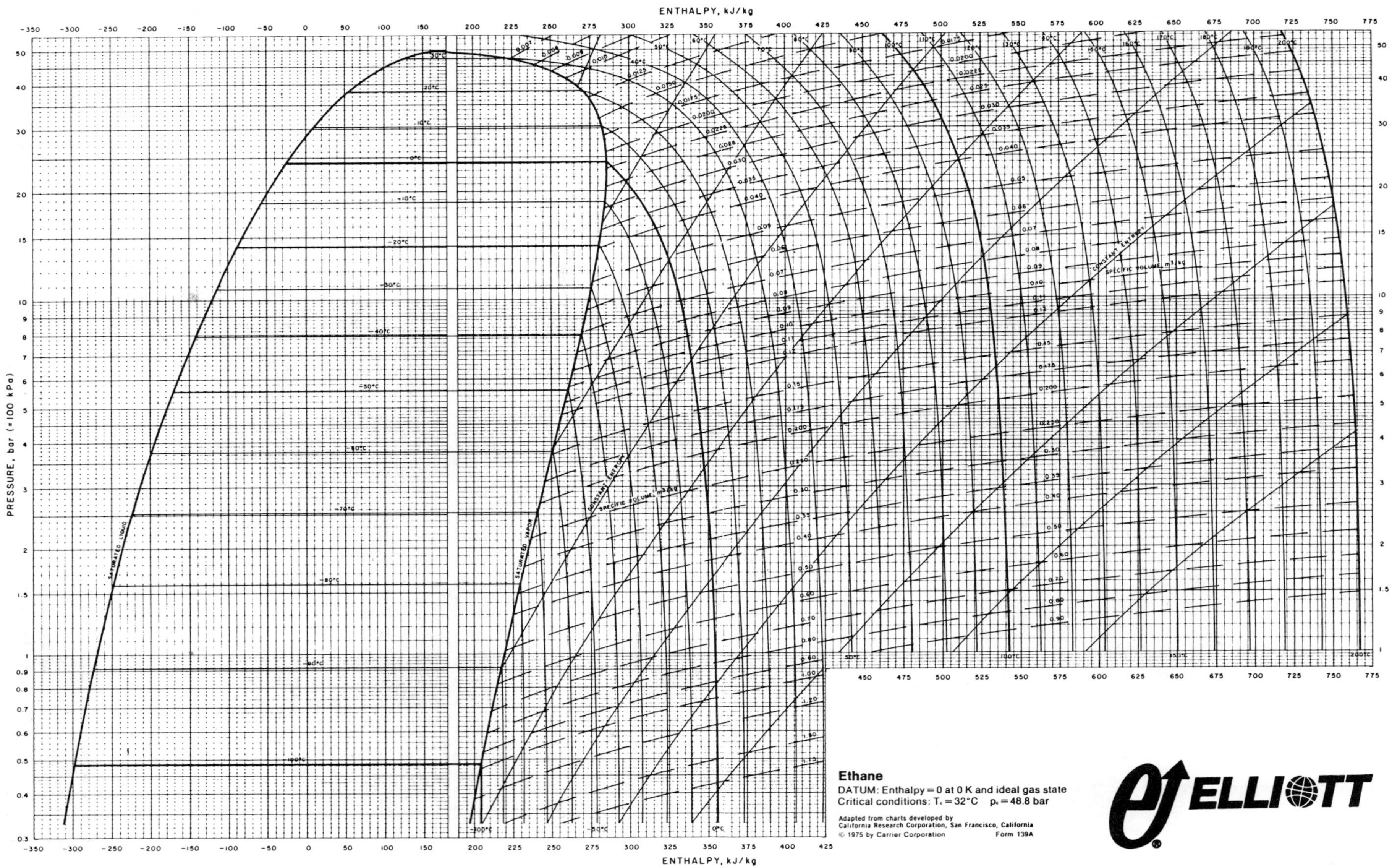

Figure B.14. Mollier diagram for ethane. *(Used with permission of Elliott Company, Jeannette, PA.)*

Ethylene

English units

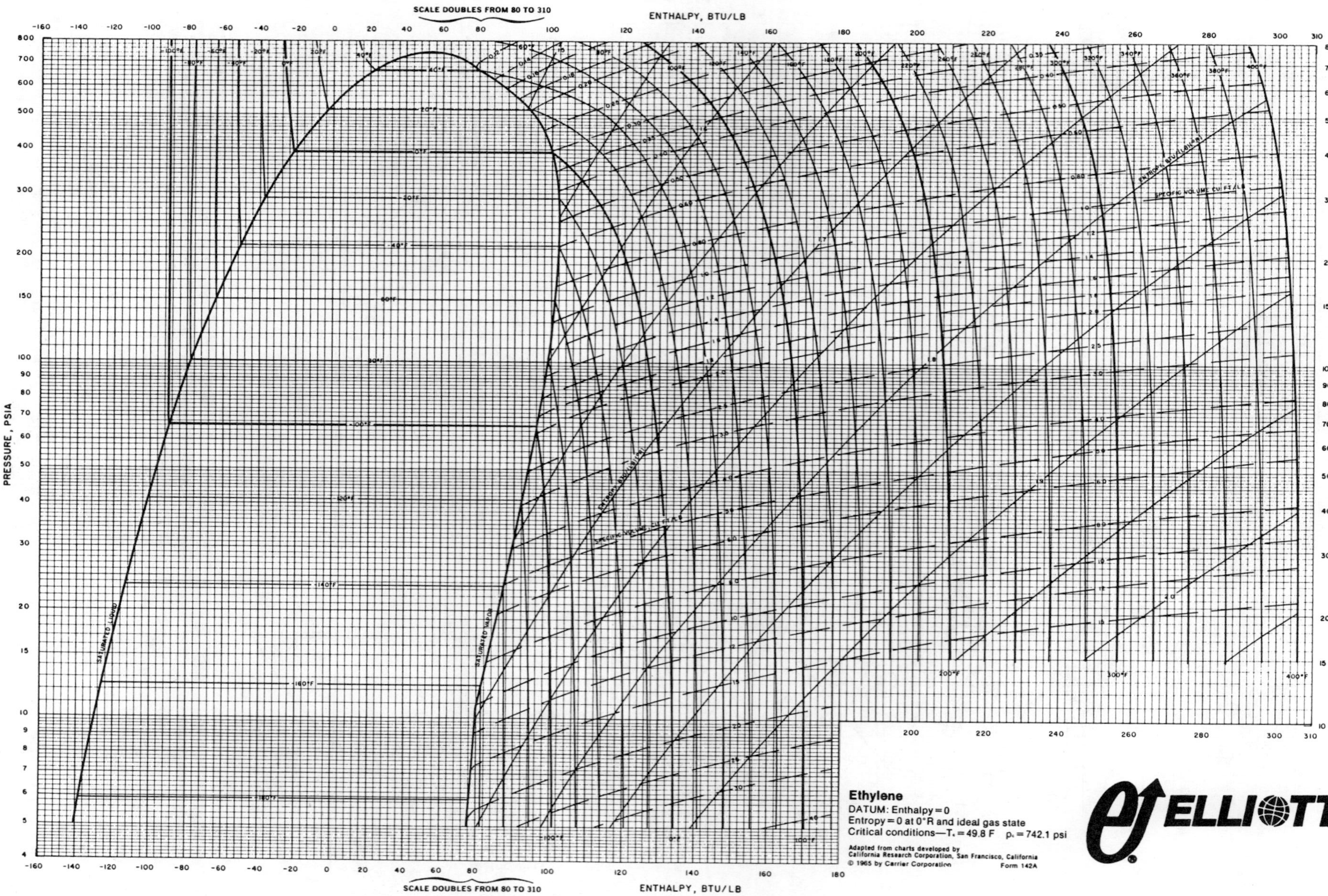

Figure B.15. Mollier diagram for ethylene. *(Used with permission of Elliott Company, Jeannette, PA.)*

Ethylene

Metric units
(see other side for English units)

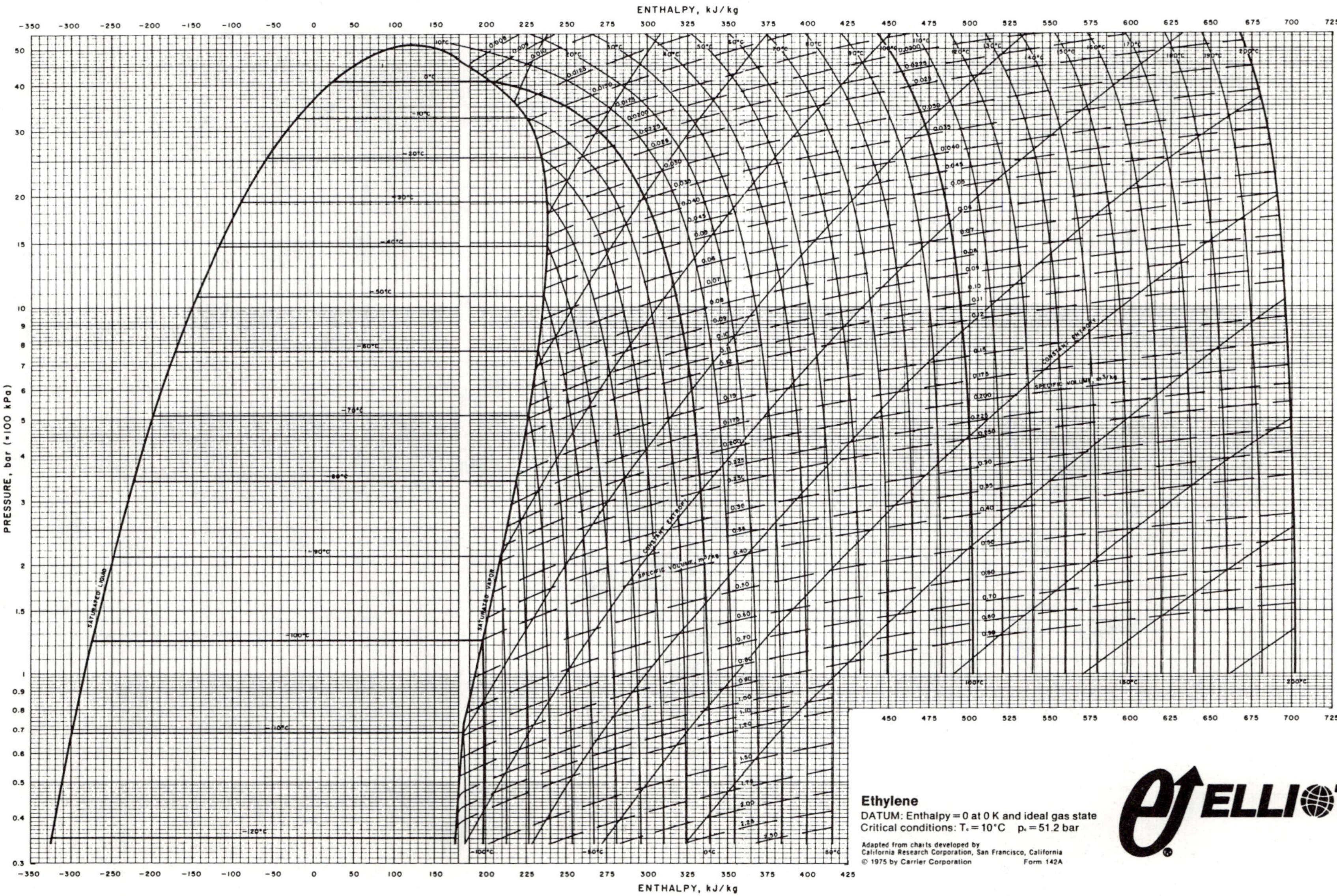

Figure B.16. Mollier diagram for ethylene. *(Used with permission of Elliott Company, Jeannette, PA.)*

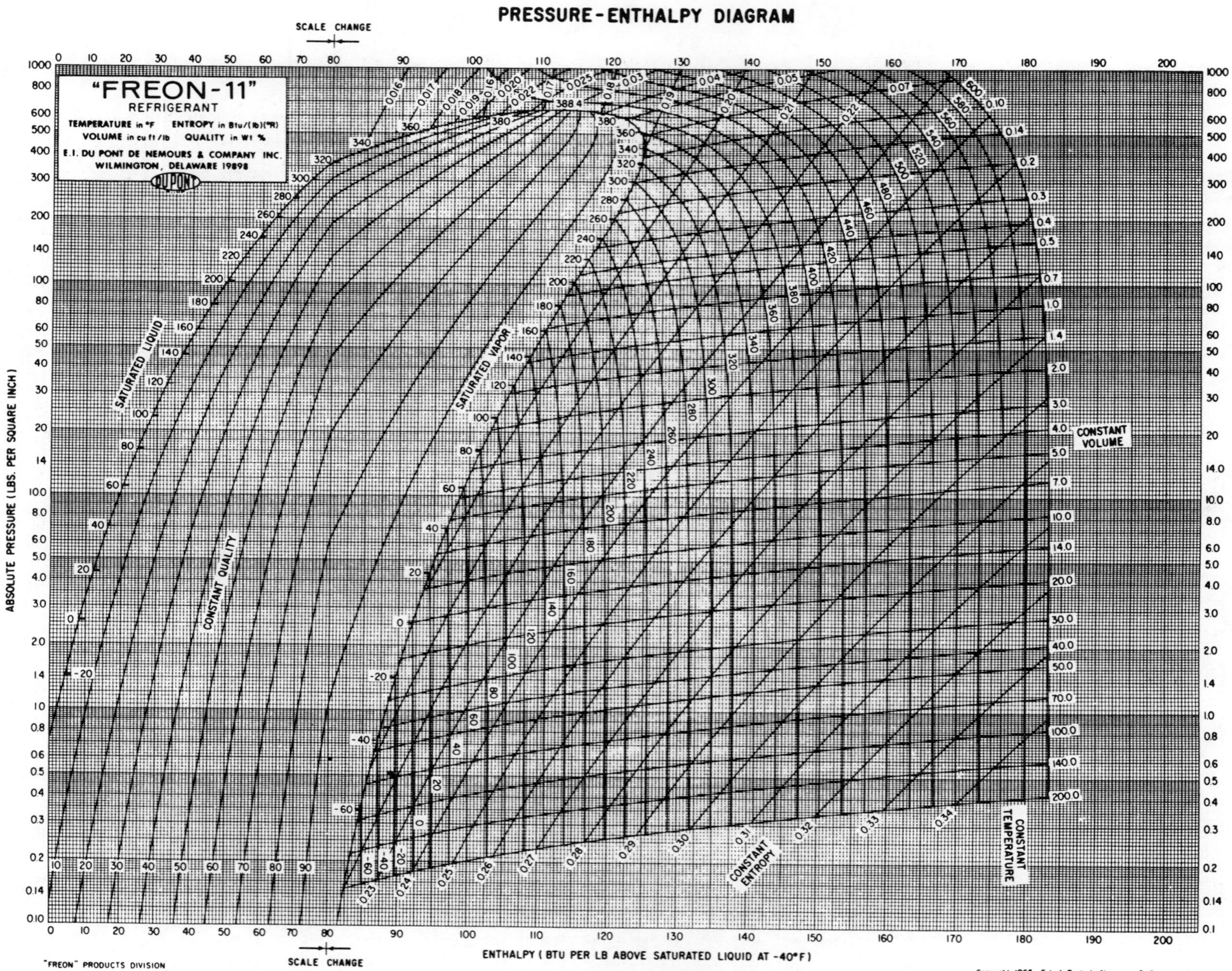

Figure B.17. Pressure-enthalpy diagram for Freon-11. *(From [31], with permission of E.I. du Pont de Nemours & Company. Copyright 1965.)*

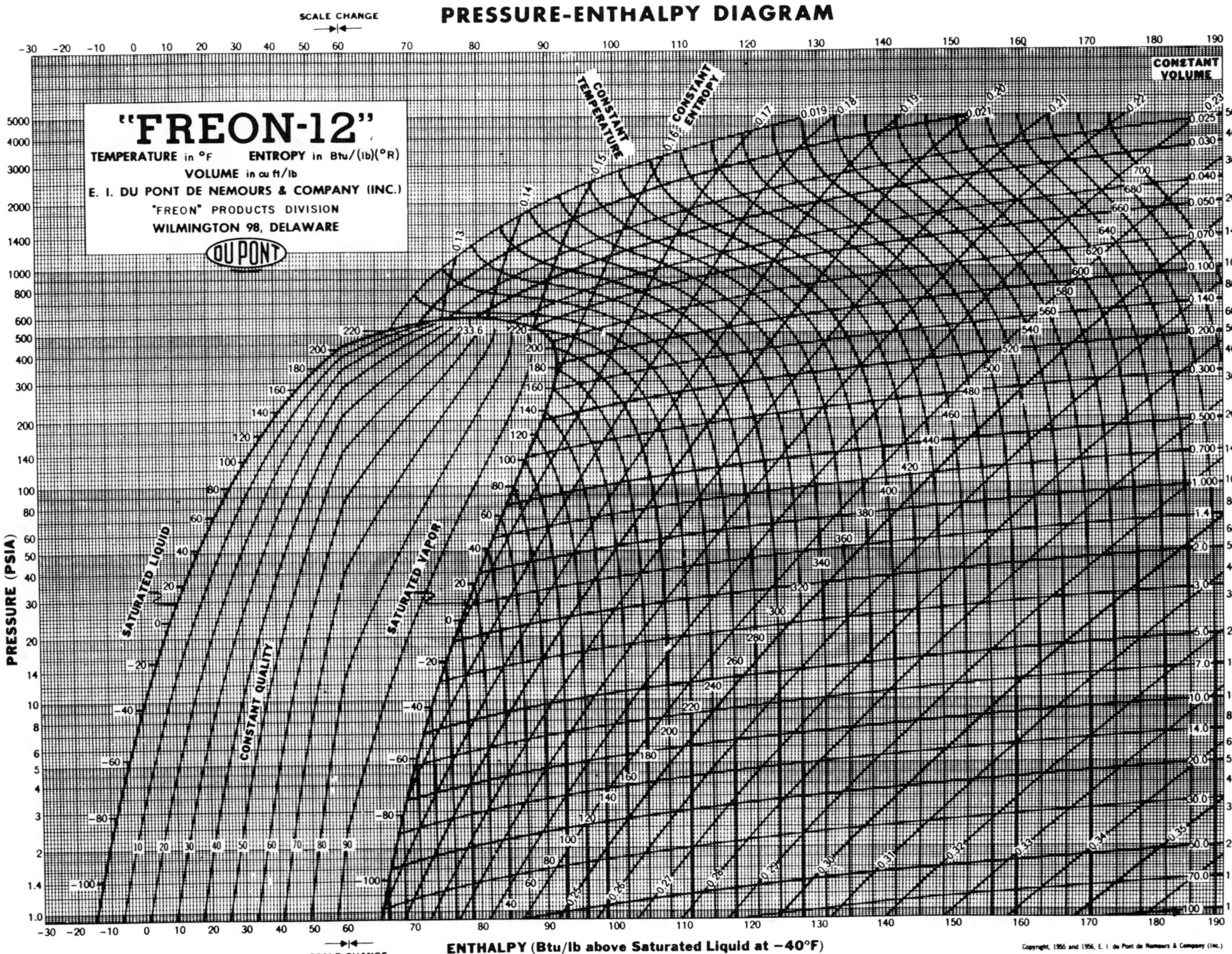

Figure B.18. Pressure-enthalpy diagram for Freon-12. *(From [31], with permission of E.I. du Pont de Nemours & Company. Copyright 1955, 1956.)*

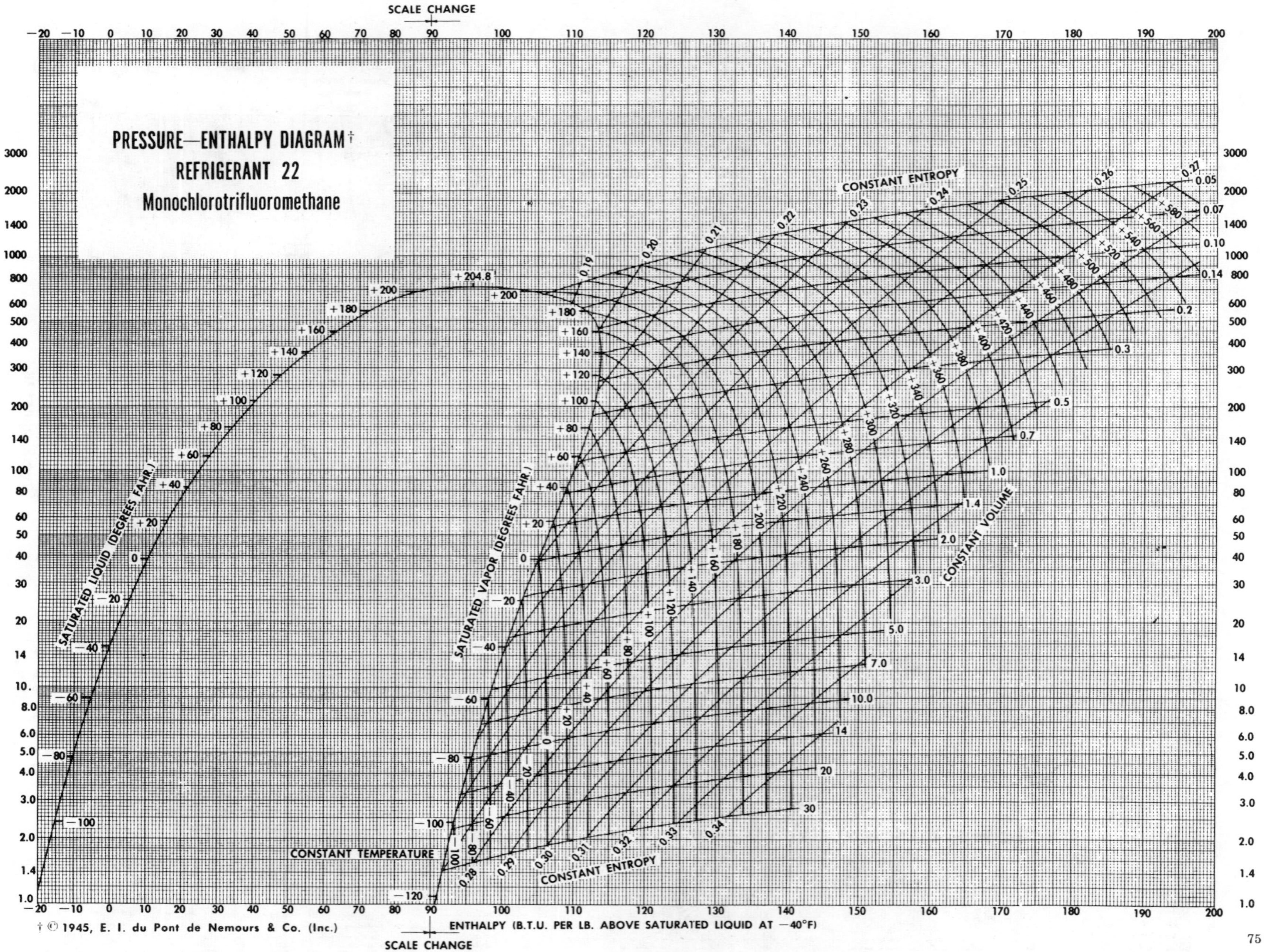

Figure B.19. Pressure-enthalpy diagram for refrigerant 22. *(From [32], with permission of E.I. du Pont de Nemours & Company. Copyright 1945.)*

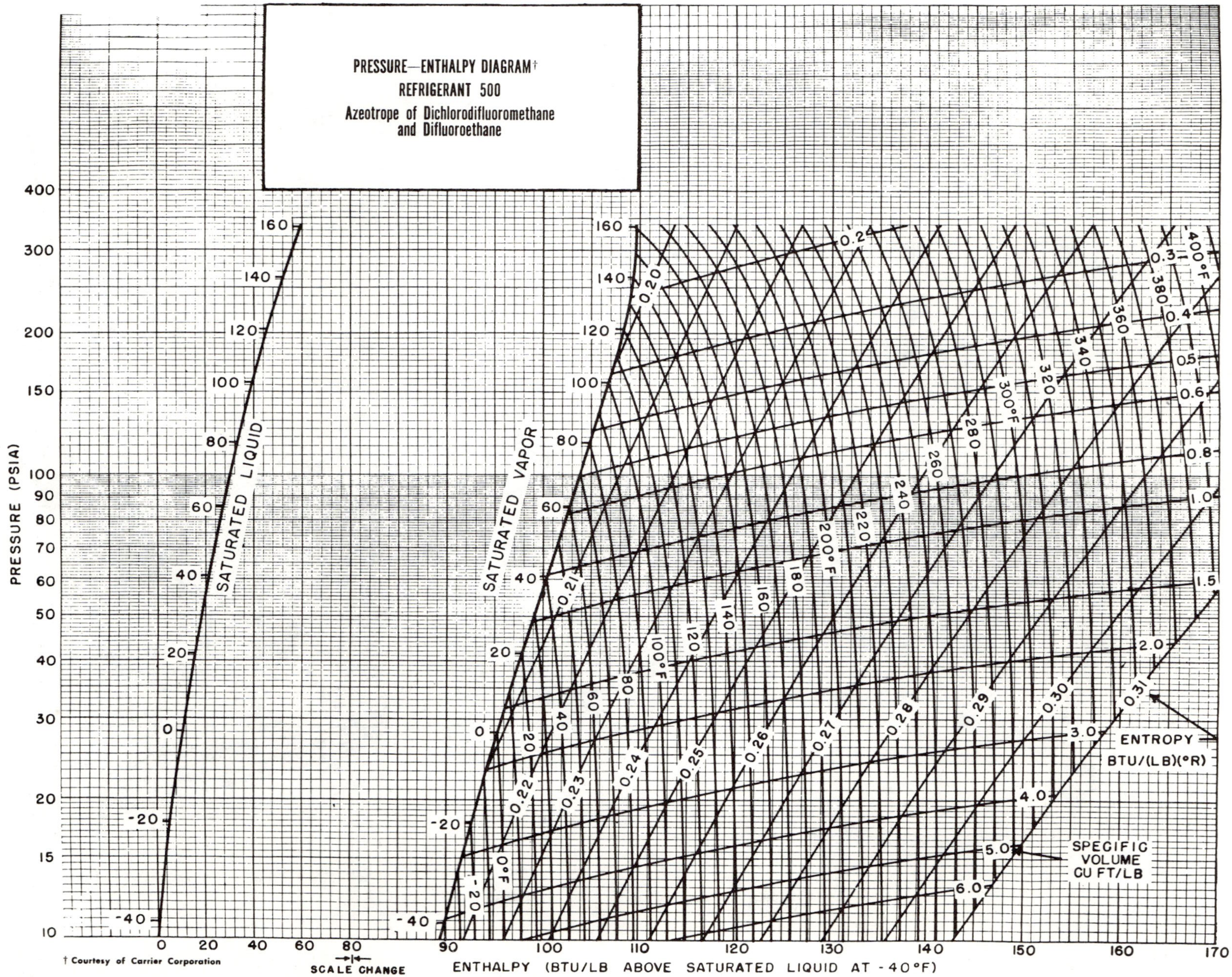

Figure B.20. Pressure-enthalpy diagram for refrigerant 500. *(From [32], courtesy of Carrier Corporation.)*

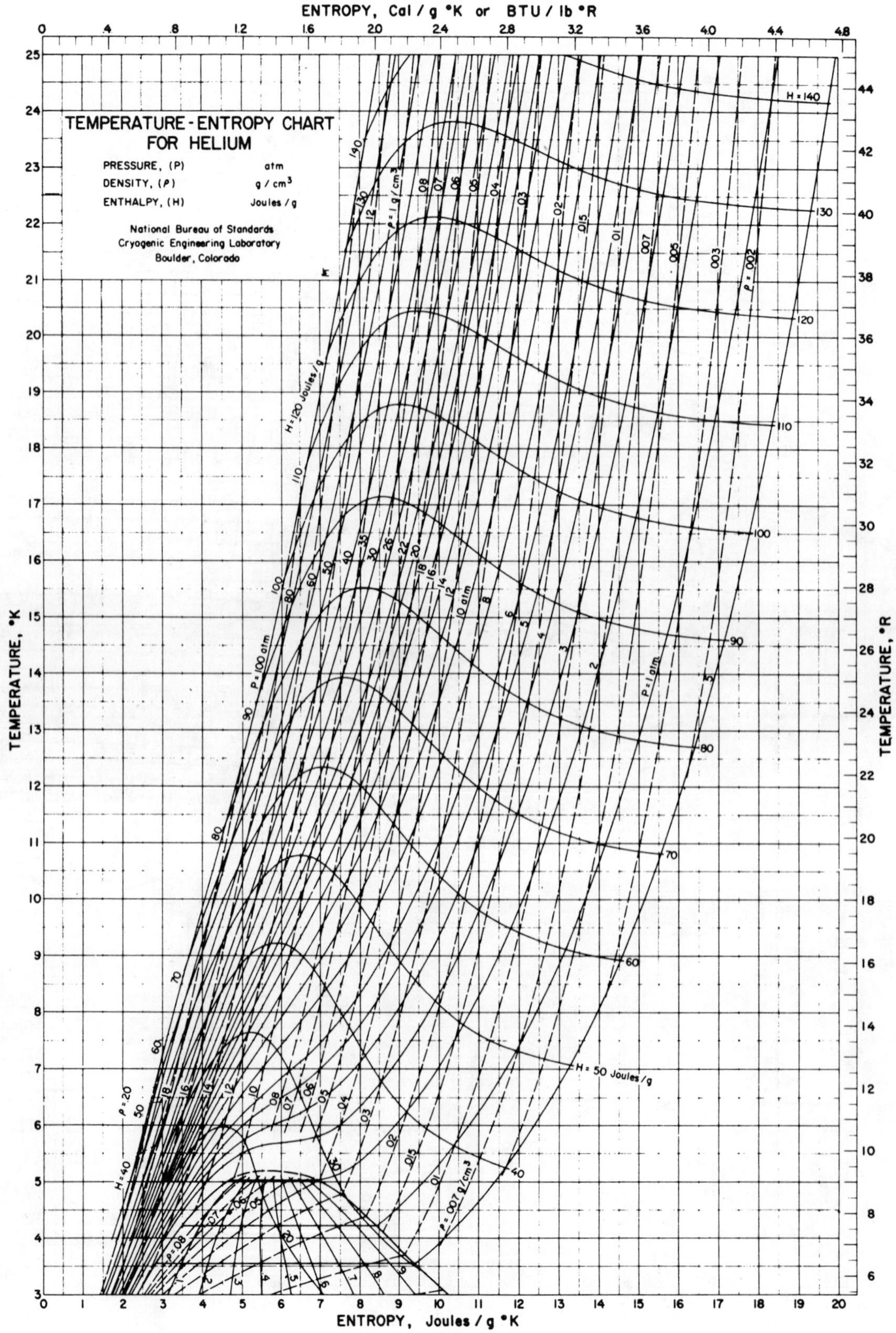

Figure B.21. Temperature-entropy diagram for helium. *(From the National Bureau of Standards, Boulder, CO.)*

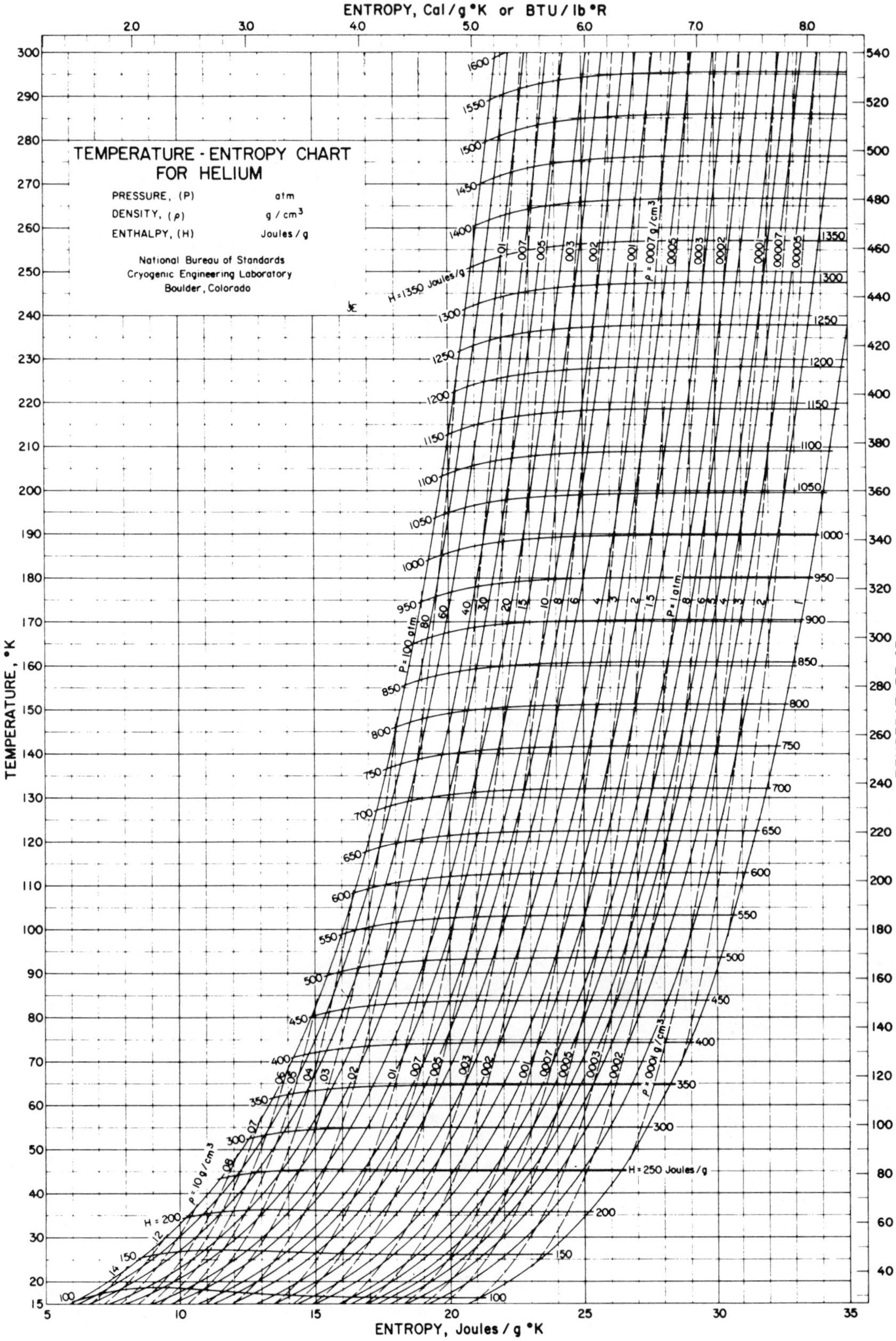

Figure B.21. (continued)

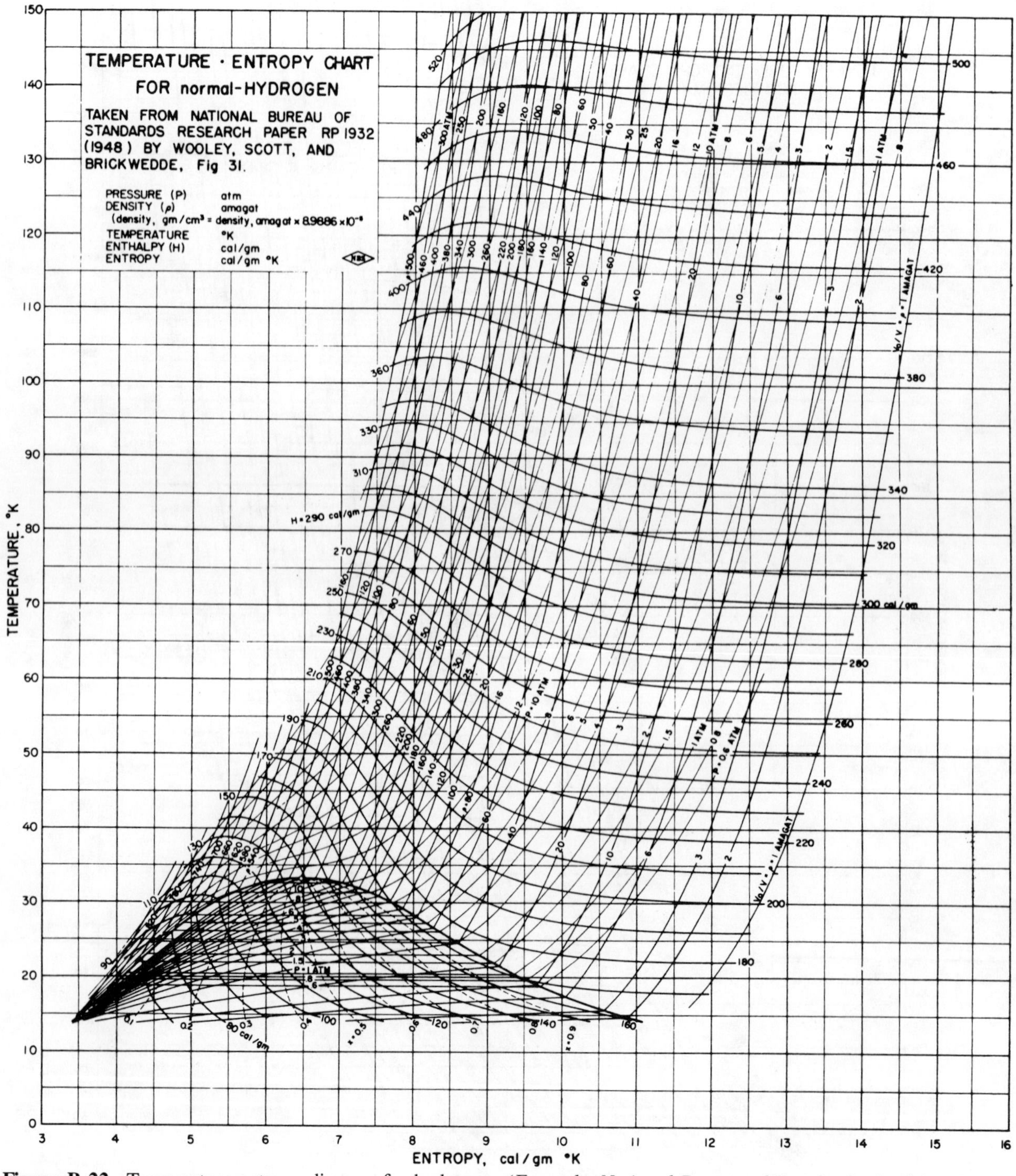

Figure B.22. Temperature-entropy diagram for hydrogen. *(From the National Bureau of Standards, Boulder, CO.)*

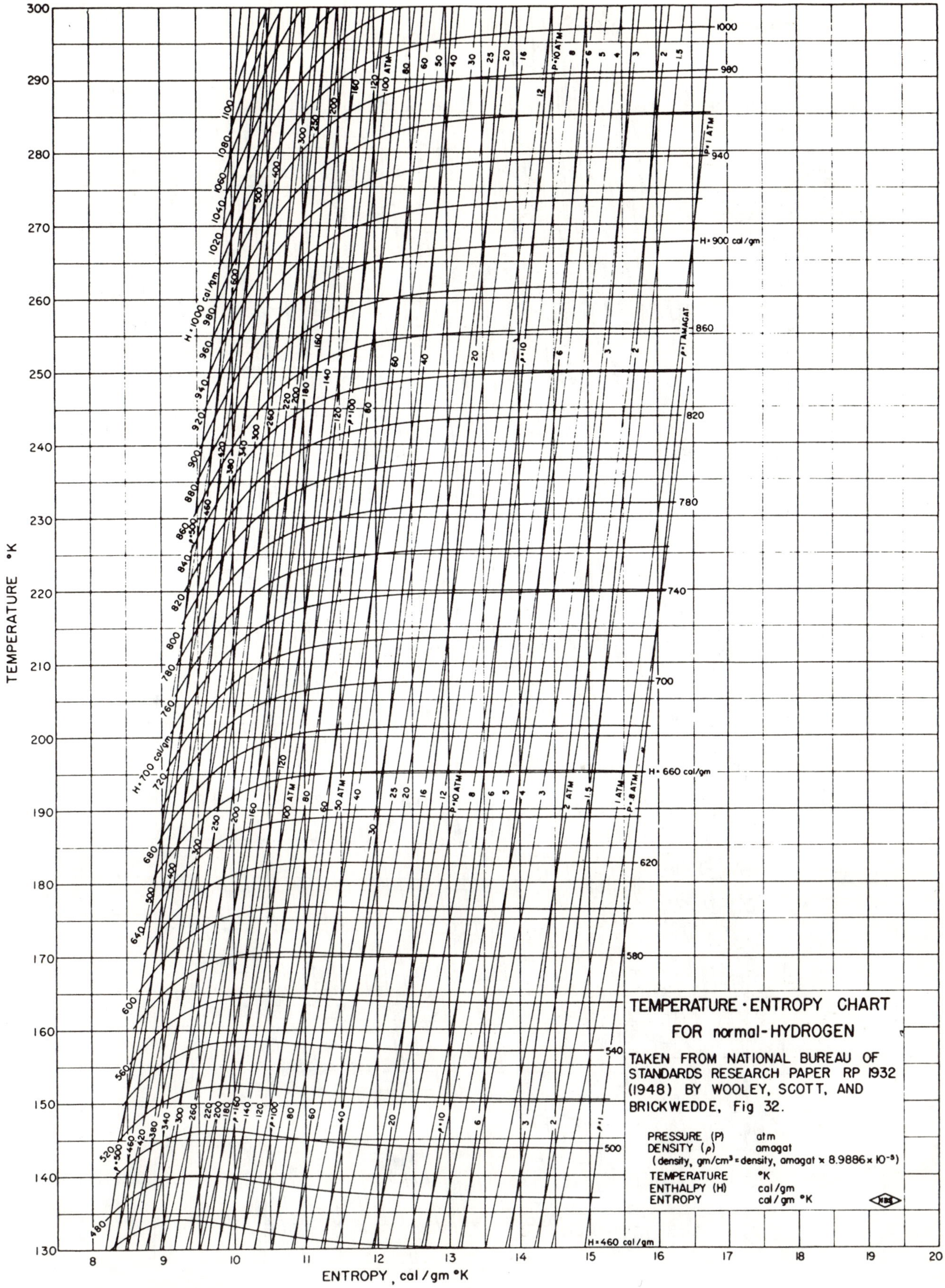

Figure B.22. (continued)

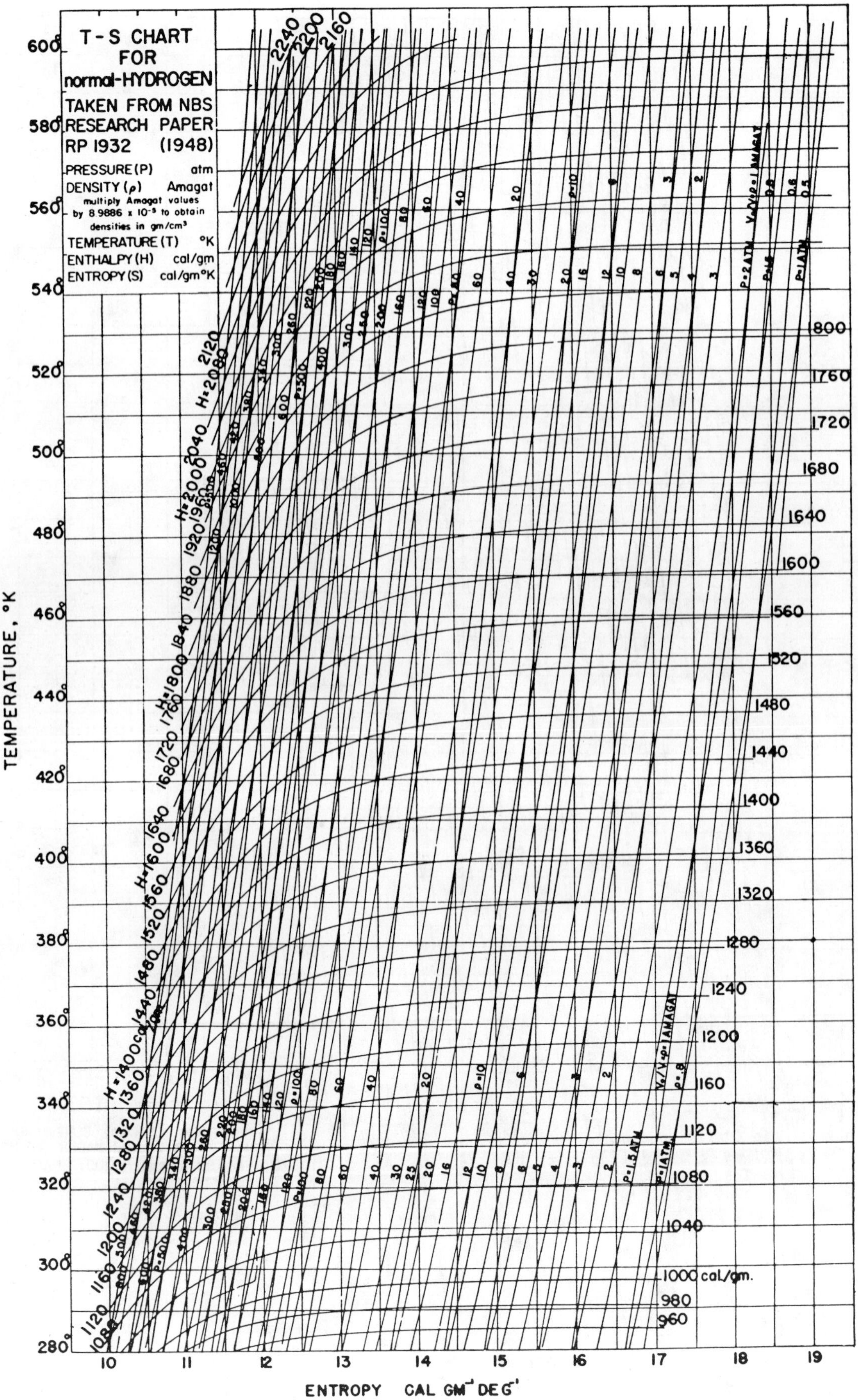

Figure B.22. (continued)

Iso-butane

English units

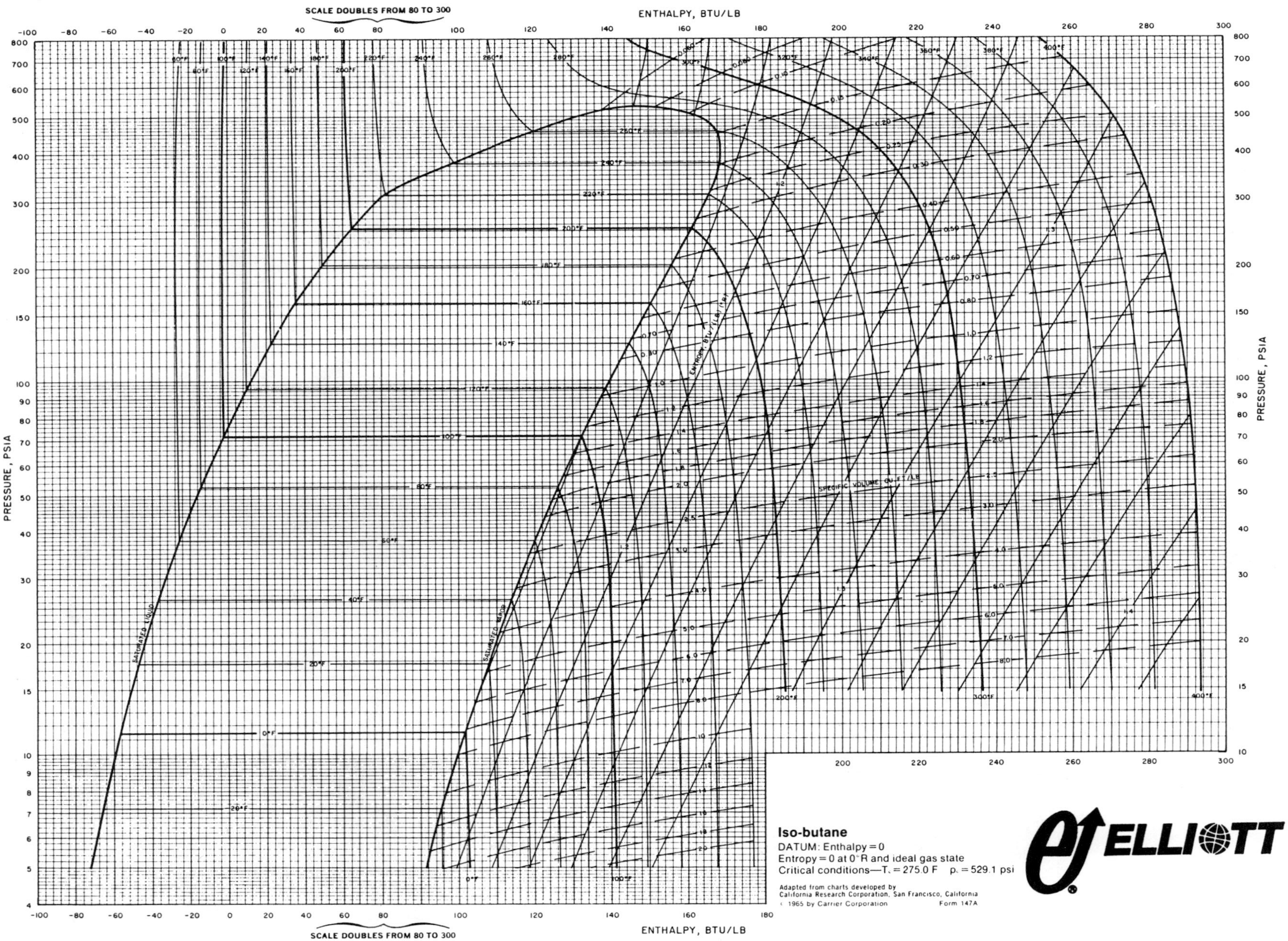

Figure B.23. Mollier diagram for iso-butane. *(Used with permission of Elliott Company, Jeannette, PA.)*

Iso-butane

Metric units

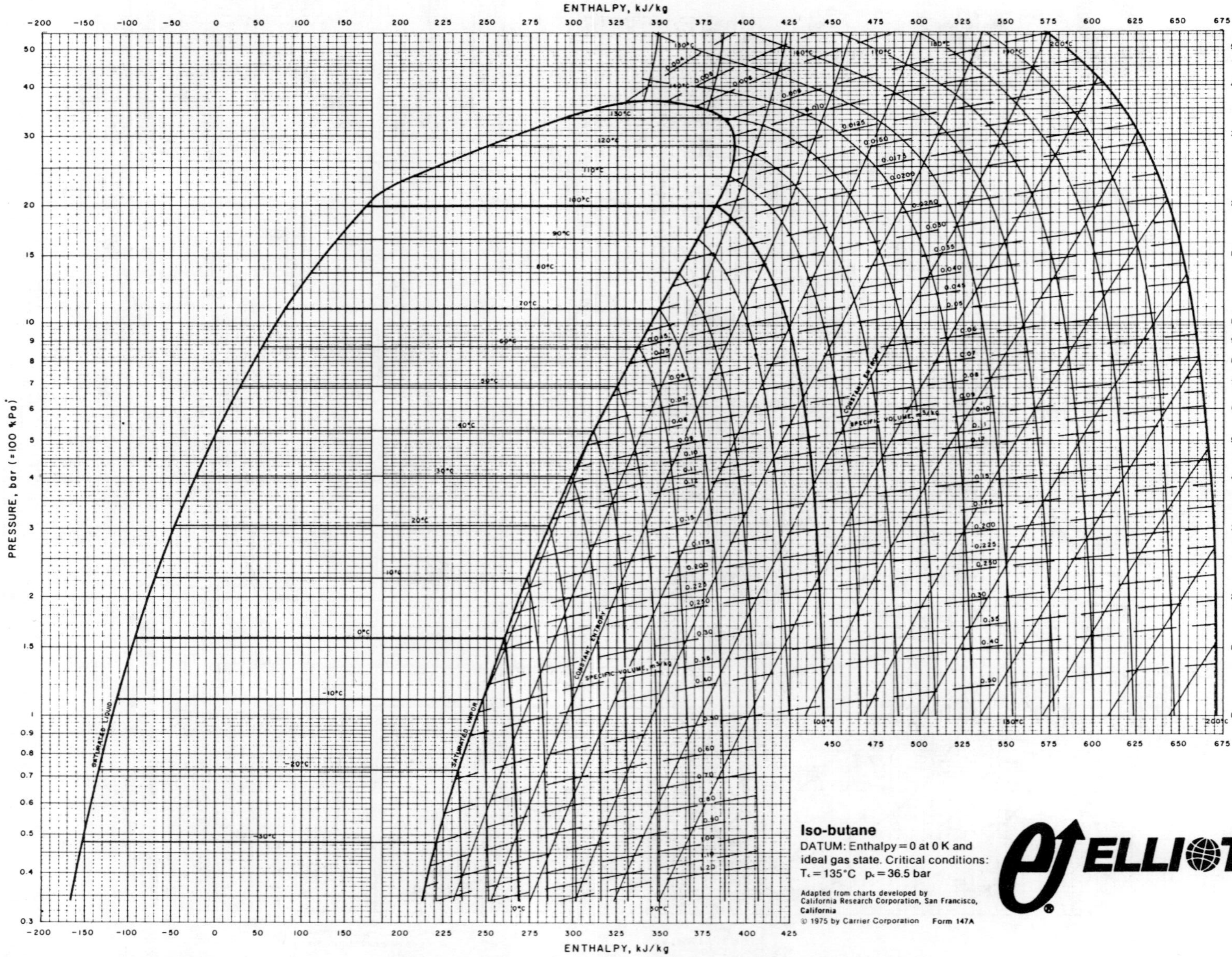

Figure B.24. Mollier diagram for iso-butane. *(Used with permission of Elliott Company, Jeannette, PA.)*

Methane

English units

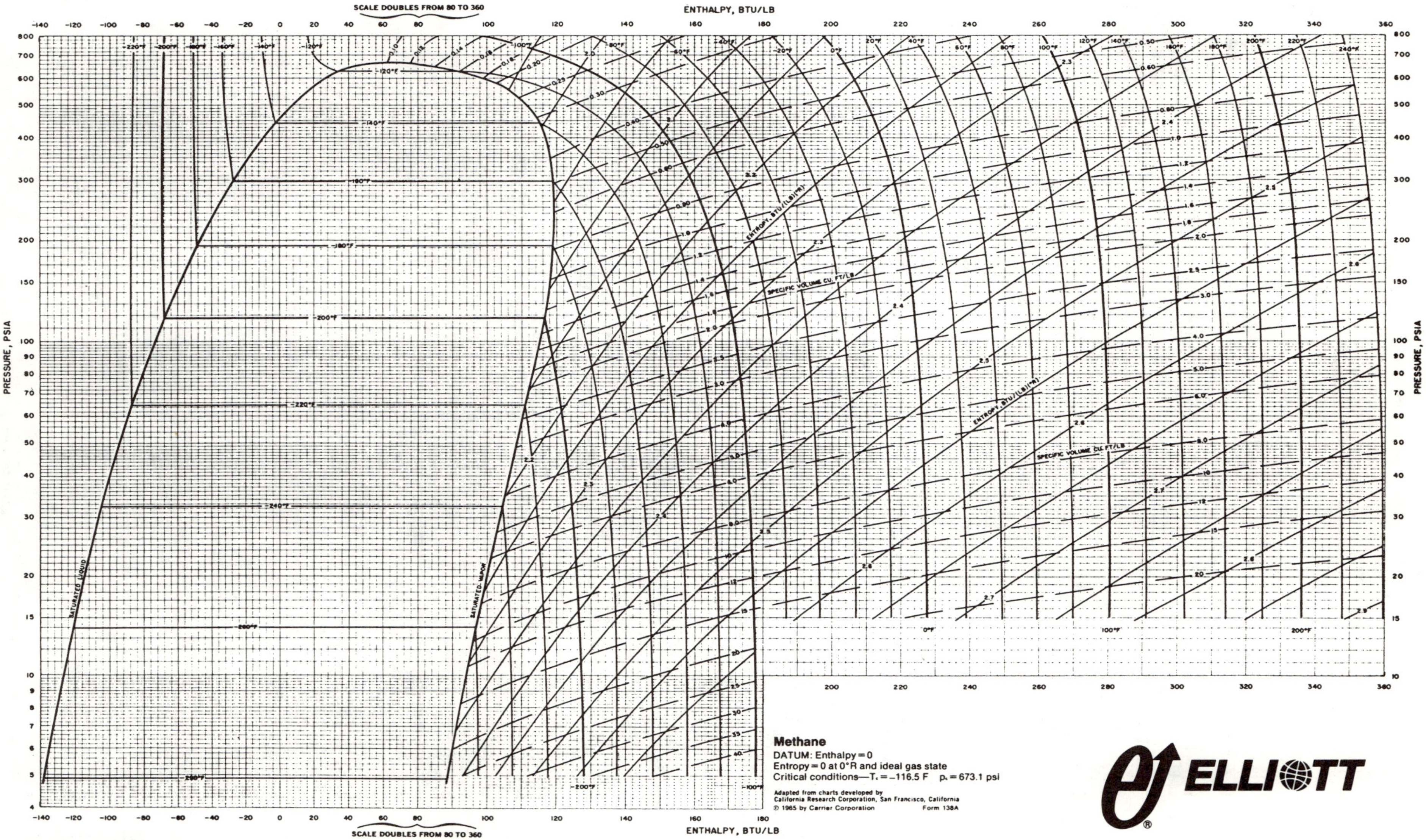

Figure B.25. Mollier diagram for methane. *(Used with permission of Elliott Company, Jeannette, PA.)*

Methane

Metric units

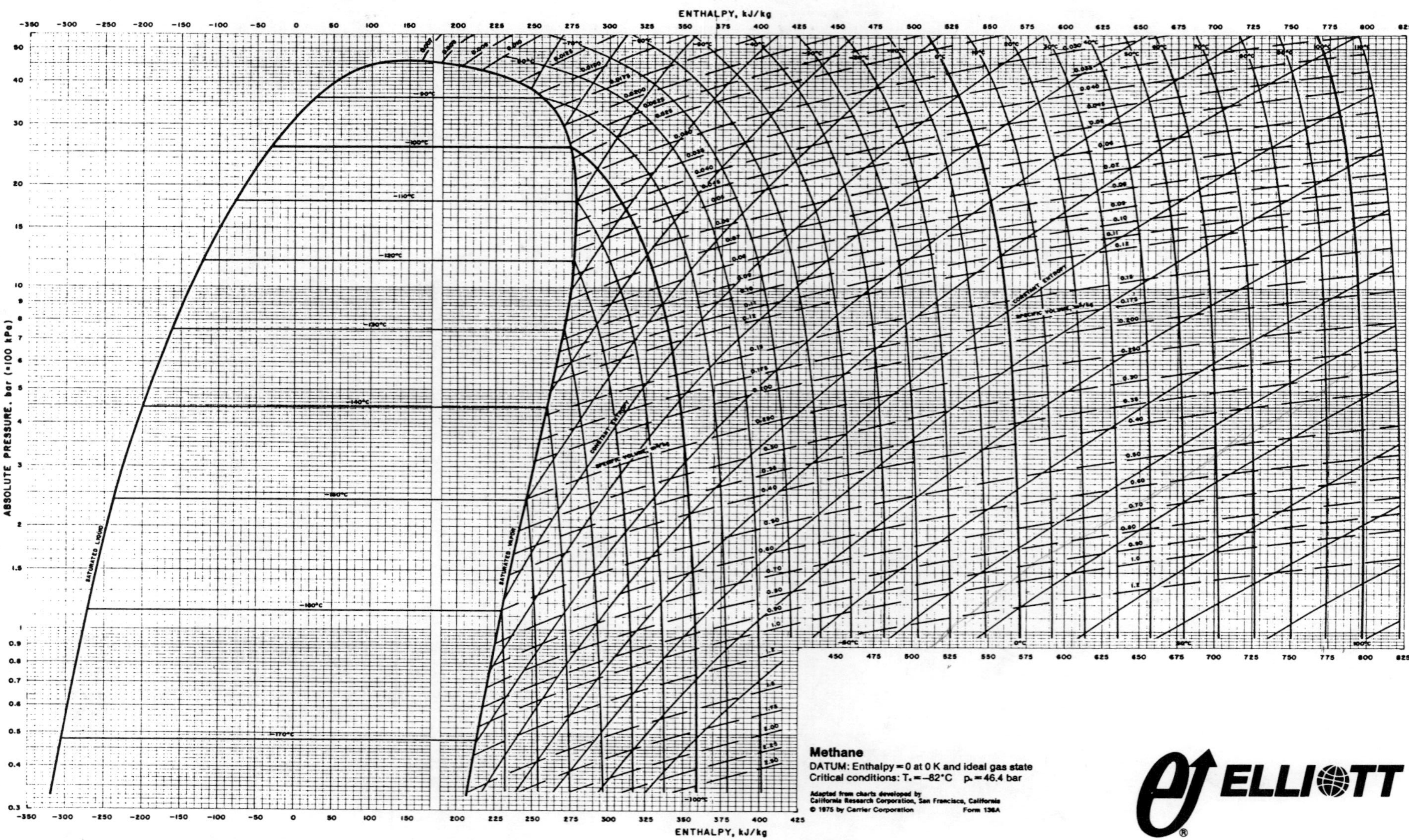

Figure B.26. Mollier diagram for methane. *(Used with permission of Elliott Company, Jeannette, PA.)*

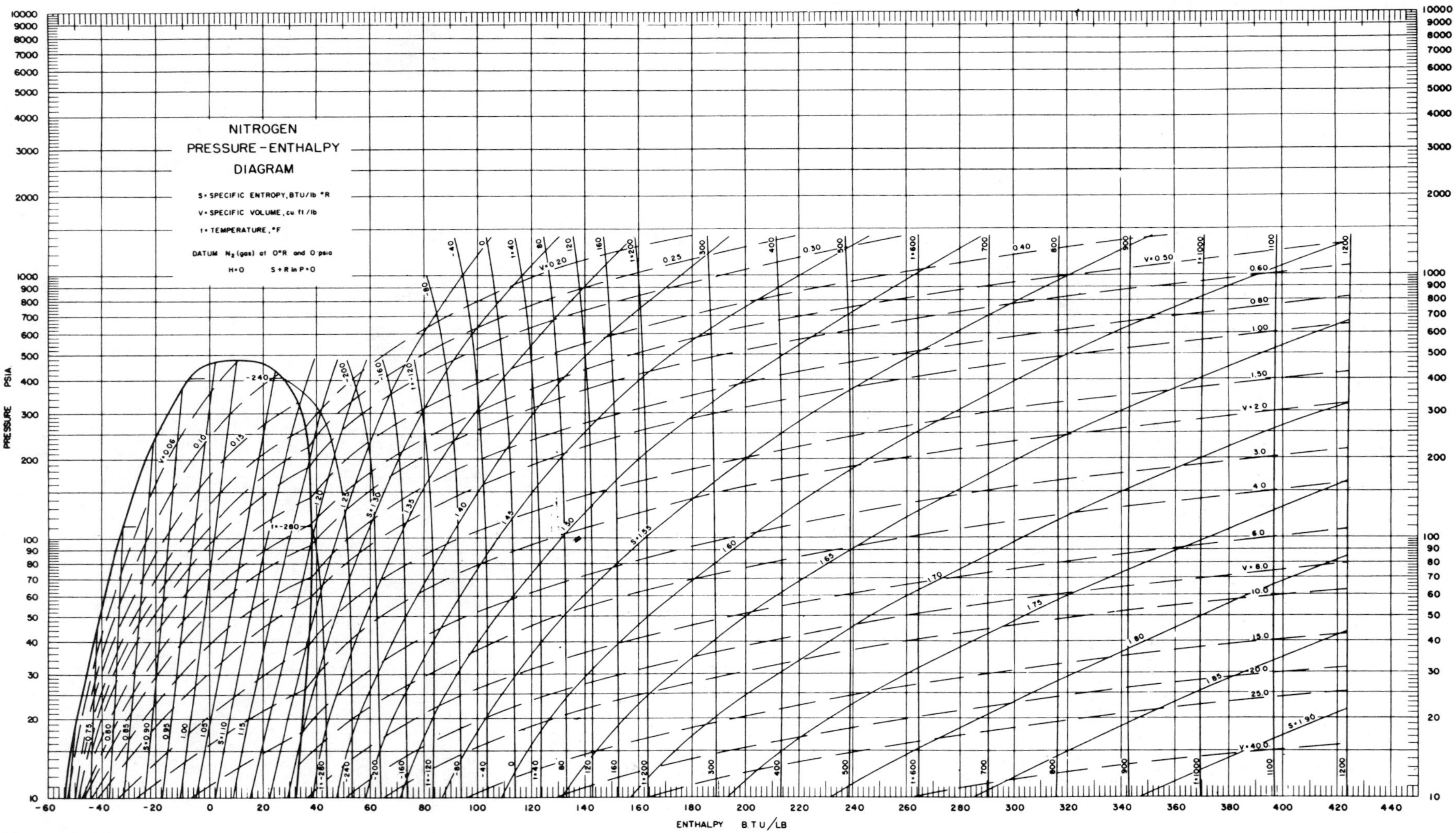

Figure B.27. Pressure-enthalpy diagram for nitrogen. *(From [31],* Thermo Properties of Non-Hydrocarbons, *by L.N. Canjar, E.K. Pollock, T.W. Cadman, W.E. Lee, and F.S. Manning. Copyright © 1966 by Gulf Publishing Company, Houston, TX. Used with permission. All rights reserved.)*

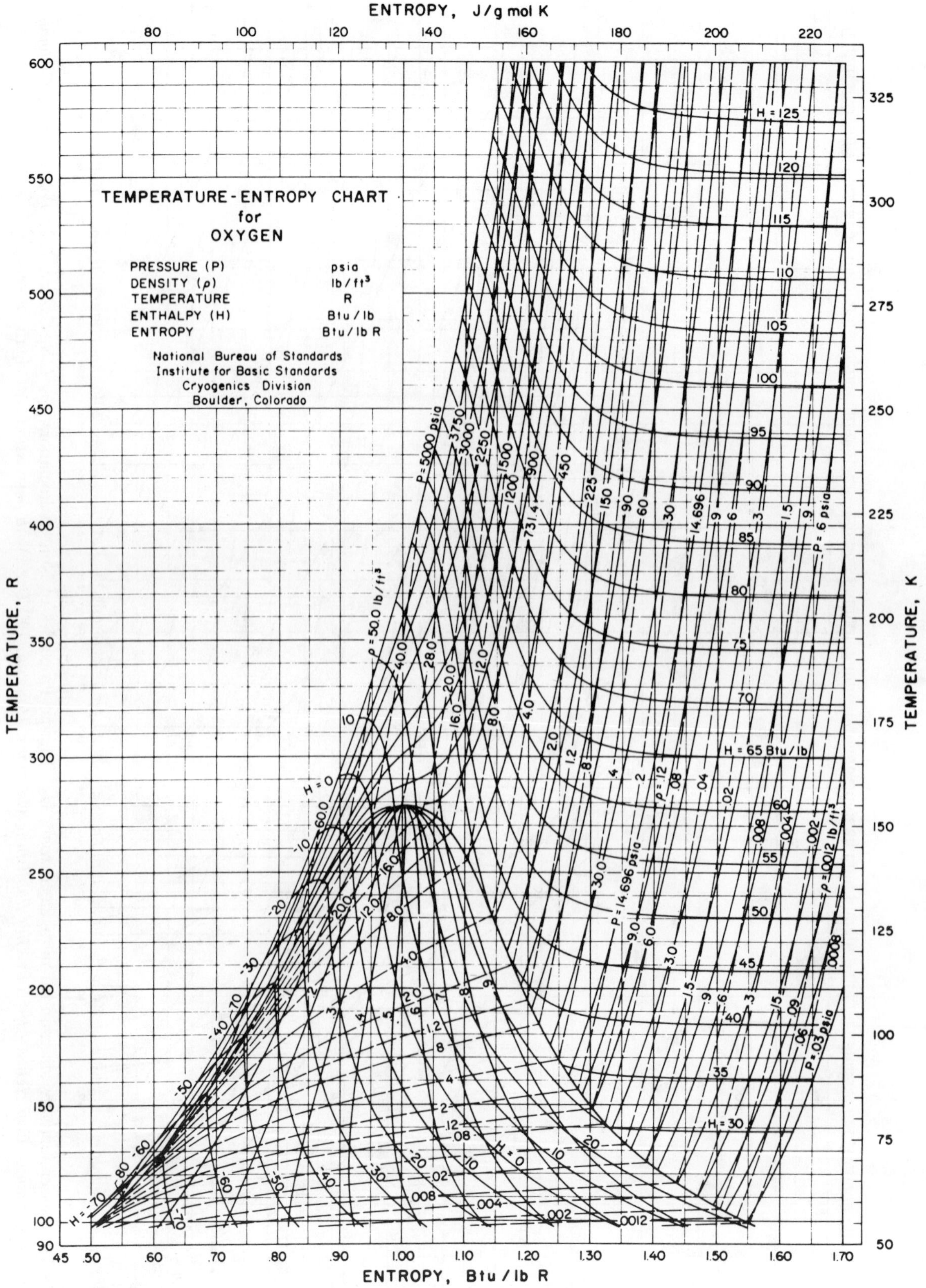

Figure B.28. Temperature-entropy diagram for oxygen. *(From National Bureau of Standards, Boulder, CO.)*

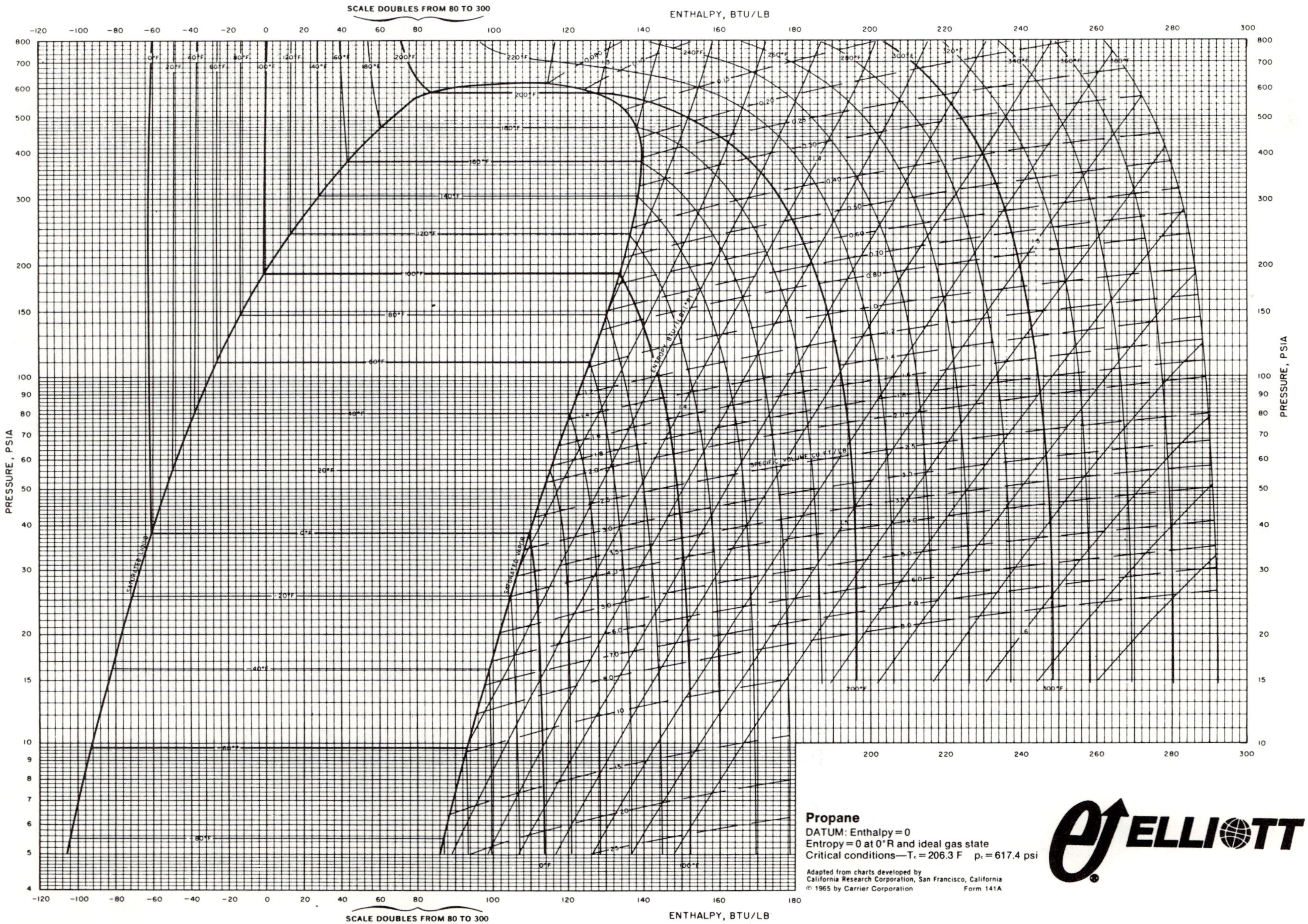

Figure B.29. Mollier diagram for propane. *(Used with permission of Elliott Company, Jeannette, PA.)*

Propane

Metric units

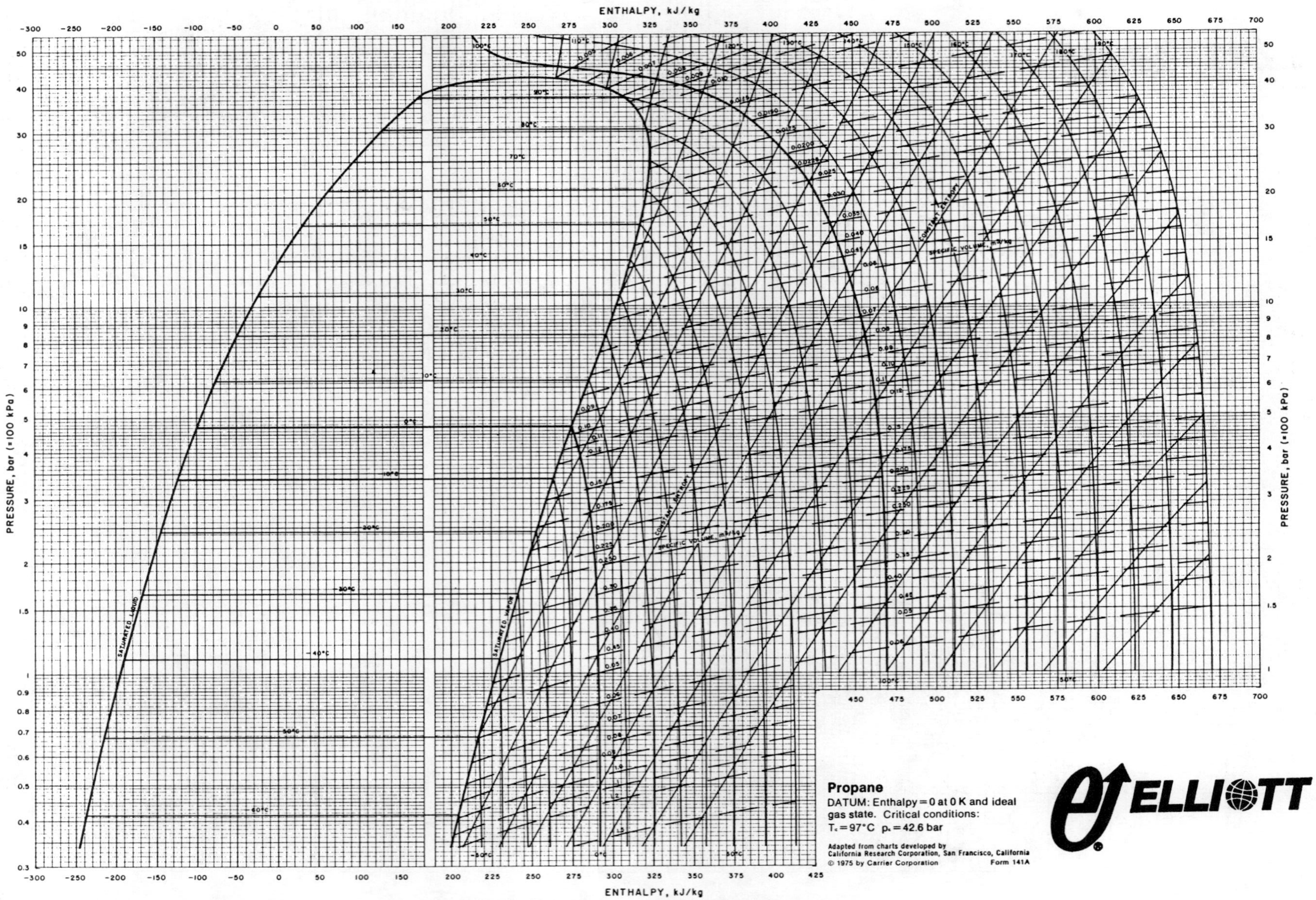

Figure B.30. Mollier diagram for propane. *(Used with permission of Elliott Company, Jeannette, PA.)*

Propylene

English units

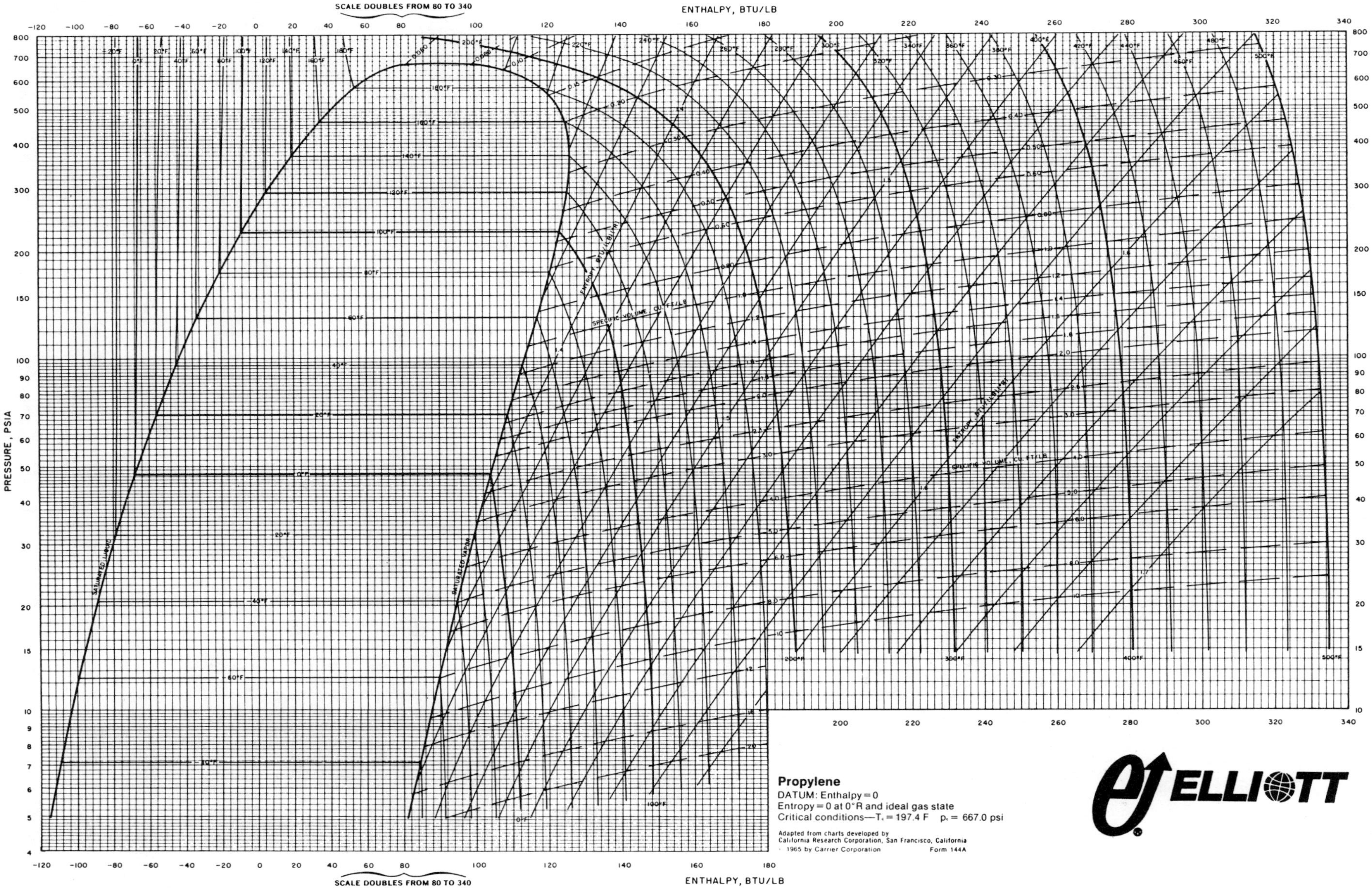

Figure B.31. Mollier diagram for propylene. *(Used with permission of Elliott Company, Jeannette, PA.)*

Propylene

Metric units

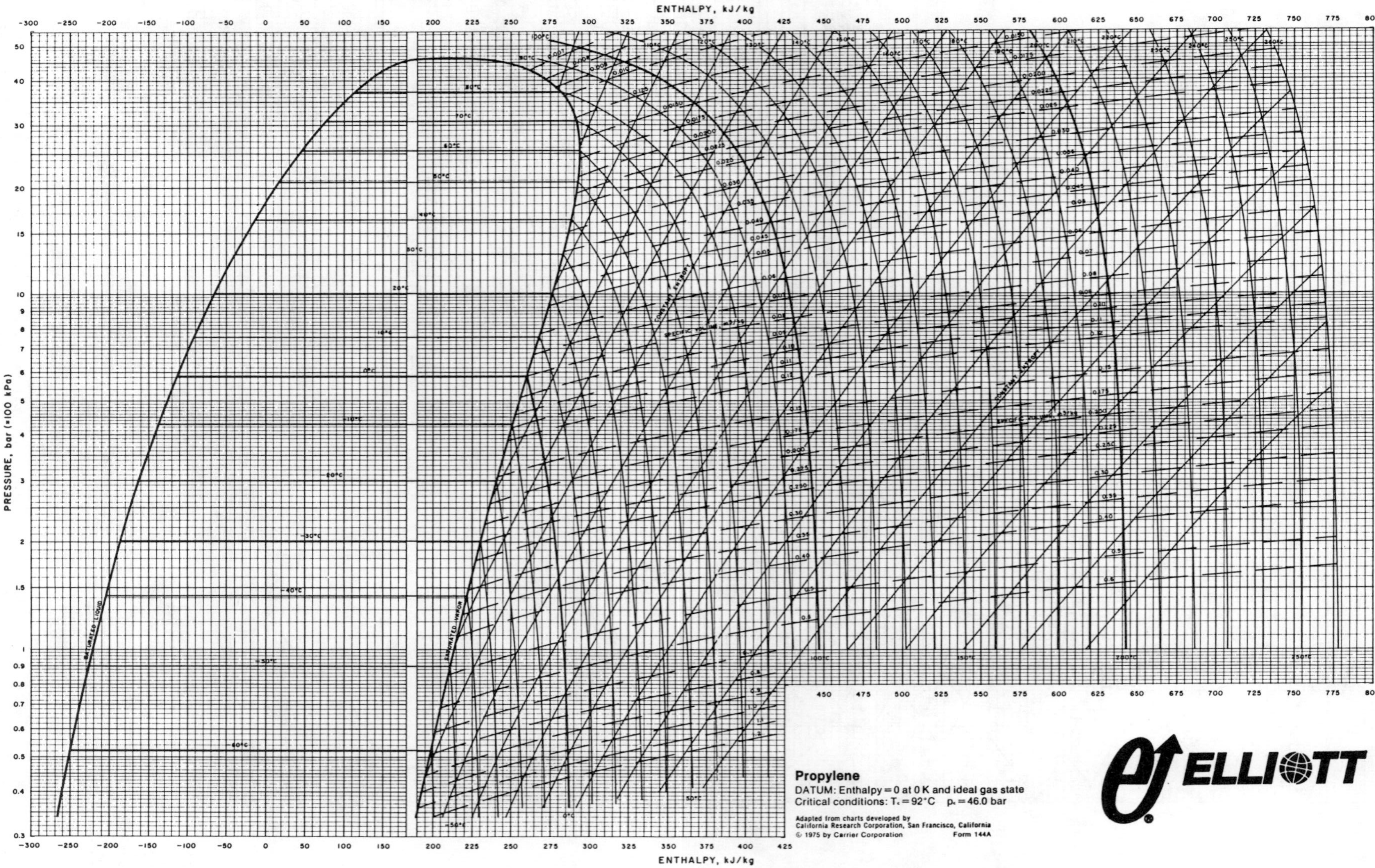

Figure B.32. Mollier diagram for propylene. *(Used with permission of Elliott Company, Jeannette, PA.)*

Adapted from fig. 22 (p. 311), 1967 ASME Steam Tables.
Copyright 1967 by the American Society of Mechanical Engineers.

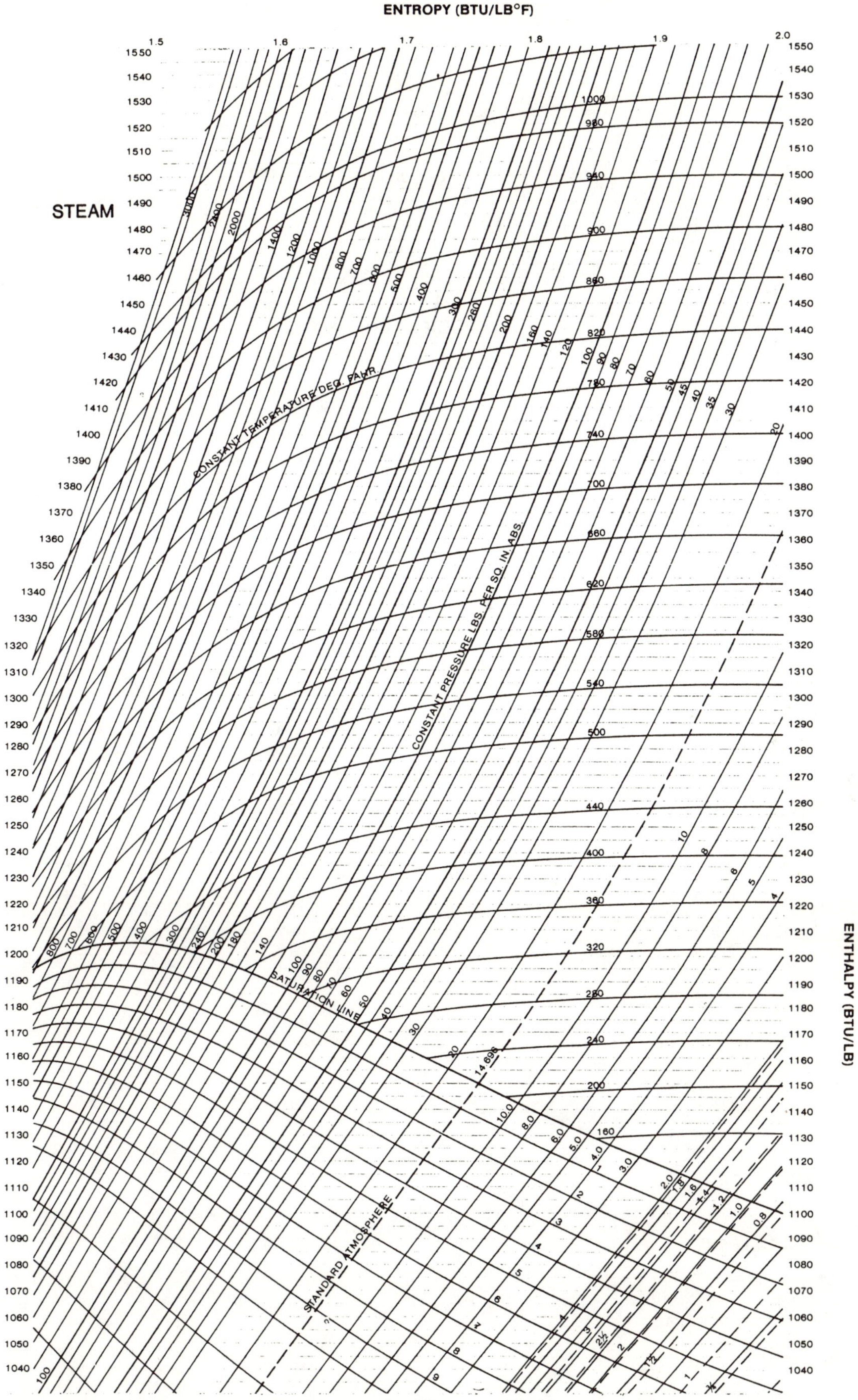

Figure B.33. Entropy-enthalpy chart for steam. *(Used with permission of Elliott Company, Jeannette, PA.)*

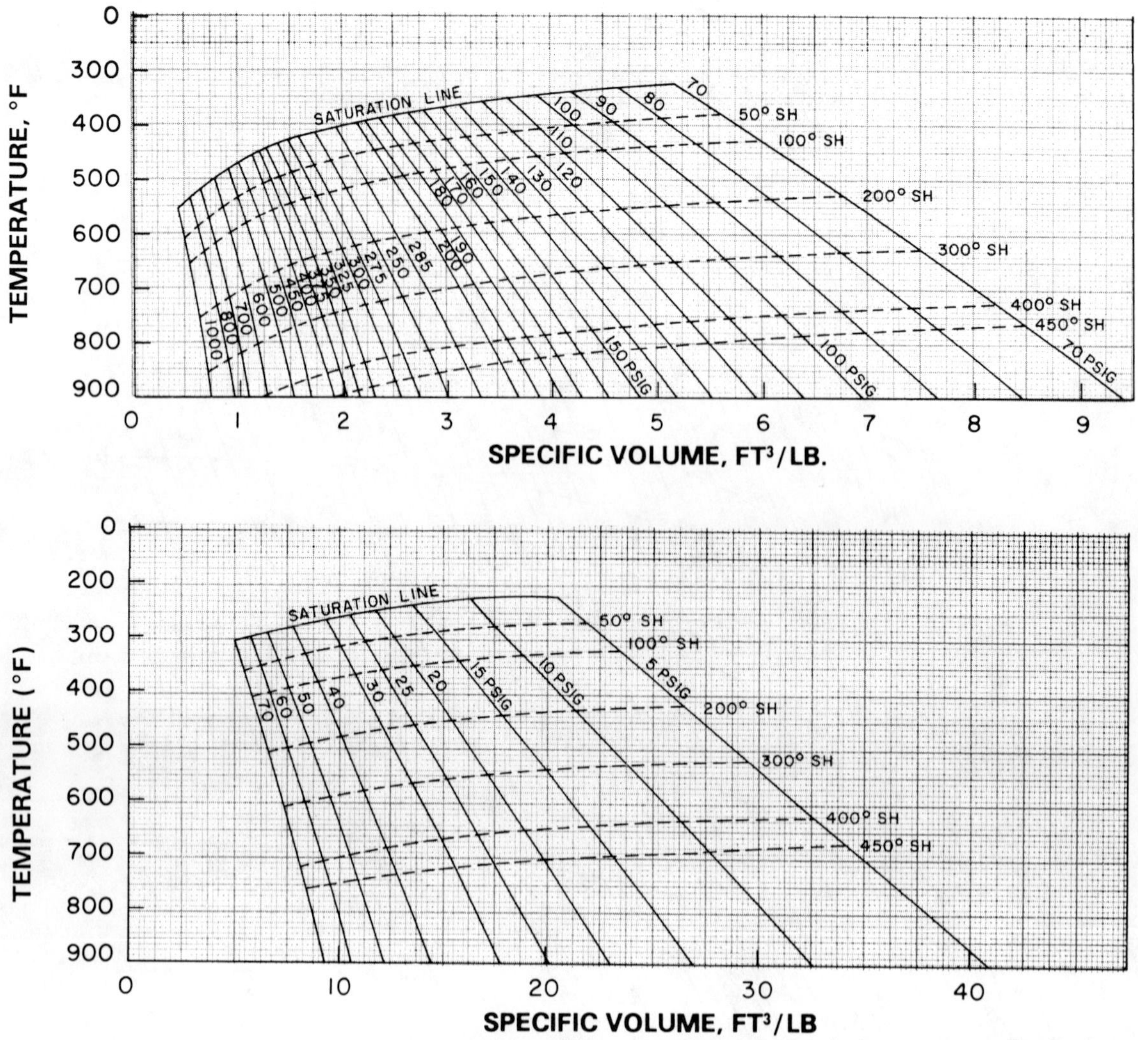

Figure B.34. Specific volume for steam. (*Used with permission of Elliott Company, Jeannette, PA.*)

Appendix C
CONVERSION TABLES

TO CONVERT	MULTIPLY BY	TO OBTAIN
	A	
acres	4.35×10^4	square feet
acres	4047.	square meters
acres	.001562	square miles
acres	4840.	square yards
ampere-hours	3600.	coulombs
ampere-hours	.03731	faradays
ampere-turns	1.257	gilberts
angstrom unit	3.937×10^{-9}	inches
angstrom unit	1×10^{-10}	meters
angstrom unit	1×10^{-4}	microns or (μ)
ares	.02471	acres (U.S.)
ares	119.6	sq. yards
ares	100	sq. meters
astronomical unit	1.495×10^8	kilometers
atmospheres	.007348	tons/in^2
atmospheres	1.058	tons/ft^2
atmospheres	76.	cm of mercury (at 0°C)
atmospheres	33.9	feet of water (at 4°C)
atmospheres	29.92	inches of mercury (at 0°C)
atmospheres	.76	meters of mercury (at 0°C)
atmospheres	760	millimeteres of mercury (at 0°C)
atmospheres	1.03323	kg/cm^2
atmospheres	1.0333×10^4	kg/m^2
atmospheres	14.7	lb/in^2
atmospheres	1.01325	bar
	B	
barrels (U.S., dry)	3.281	bushels
barrels (U.S., dry)	7056.	cu. inches
barrels (U.S., dry)	105.	quarts (dry)
barrels (U.S., liquid)	31.5	gallons
barrels (oil)	42.	gallons (oil)
bars	.986923	atmospheres
bars	1×10^6	dynes/cm^2
bars	1.020×10^4	kg/m^2
bars	2089.	lb/ft^2
bars	14.5038	lb/in^2
bars	1.01972	kg/cm^2
bars	750.062	millimeters of mercury (at 0°C)
bars	29.5300	inches of mercury (at 32°F)
bars	33.488	ft water (at 60°F)
barye	1.00	dynes/cm^2
bolt (U.S., cloth)	36.576	meters
BTU	10.409	liter-atmospheres
BTU	1.0550×10^{10}	ergs
BTU	778.16	foot-pounds

TO CONVERT	MULTIPLY BY	TO OBTAIN
BTU	252.	gram-calories
BTU	3.93011×10^{-4}	horsepower-hours
BTU	1055.05	joules
BTU	.252	kilogram-calories
BTU	107.585	kilogram-meters
BTU	2.928×10^{-4}	kilowatt-hours
BTU/hr	.2162	ft-lb/sec
BTU/hr	.070	g-cal/sec
BTU/hr	3.929×10^{-4}	horsepower
BTU/hr	.2931	watts
BTU/min	12.96	ft-lb/sec
BTU/min	.02356	horsepower
BTU/min	.01757	kilowatts
BTU/min	17.57	watts
BTU/lb_m	2.326	kilojoules/kilogram
BTU/lb_m	2.326	kilonewton-meters/kilograms
BTU/ft^2/min	.0122	watts/in^2
bucket (Br. dry)	1.8184×10^4	cubic cm
bushels	1.2445	cubic ft
bushels	2150.4	cubic in
bushels	.03524	cubic meters
bushels	34.24	liters
bushels	4.0	pecks
bushels	64.	pints (dry)
bushels	32.	quarts (dry)

C

TO CONVERT	MULTIPLY BY	TO OBTAIN
calories, gram (mean)	.0039685	BTU (mean)
candle/cm^2	3.146	lamberts
candle/in^2	.4870	lamberts
centares	1.0	m^2
centigrade (degrees)	$(°C \times 9/5) + 32$	°F
centigrade (degrees)	$°C + 273.18$	°K
centigrams	.01	grams
centiliters	.3382	ounce (fluid) U.S.
centiliters	.6103	cubic inch
centiliters	2.705	drams
centiliters	.010	liters
centimeters	.03281	feet
centimeters	.3937	inches
centimeters	$1. \times 10^{-5}$	kilometers
centimeters	.01	meters
centimeters	6.214×10^{-6}	miles
centimeters	10.	millimeters
centimeters	393.7	mils
centimeters	.001094	yards
centimeters	$1. \times 10^4$	microns
centimeters	$1. \times 10^8$	angstrom units
centimeter-dynes	.001020	cn-grams
centimeter-dynes	1.020×10^{-8}	meter-kgs
centimeter-dynes	7.376×10^{-8}	lb-ft

TO CONVERT	MULTIPLY BY	TO OBTAIN
centimeter-grams	980.7	cm-dynes
centimeter-grams	$1.\times10^{-5}$	meter-kg
centimeter-grams	7.233×10^{-5}	pound-ft
centimeters of mercury	.01316	atmospheres
centimeters of mercury	.4461	ft of water
centimeters of mercury	136.	kgs/m^2
centimeters of mercury	27.85	lb/ft^2
centimeters of mercury	.1934	lb/in^2
centimeters/sec	1.969	ft/min
centimeters/sec	.03281	ft/sec
centimeters/sec	.036	km/hr
centimeters/sec	.01934	knots
centimeters/sec	.60	meters/min
centimeters/sec	.02237	miles/hr
centimeters/sec/sec	.03281	ft/sec/sec
centimeters/sec/sec	.036	km/hr/sec
centimeters/sec/sec	.010	meters/sec/sec
centimeters/sec/sec	.02237	miles/hr/sec
centipoise	.010	gr/cm/sec
centipoise	6.27×10^{-4}	lb/ft-sec
centipoise	2.4	lb/ft-hr
circumference	6.283	radians
cords	8.0	cord ft
cord ft	16.	cubic ft
coulombs	2.998×10^{9}	statcoulombs
coulombs	1.036×10^{-5}	faradays
$coulombs/cm^2$	6.452	$coulombs/in^2$
$coulombs/cm^2$	1.0×10^{4}	$coulombs/m^2$
$coulombs/in^2$	.1550	$coulombs/cm^2$
$coulombs/in^2$	1550.	$coulombs/m^2$
$coulombs/m^2$	1.0×10^{-4}	$coulombs/cm^2$
$coulombs/m^2$	6.452×10^{-4}	$coulombs/in^2$
cubic centimeters	$3.531.\times10^{-5}$	cubic ft
cubic centimeters	.06102	cubic in
cubic centimeters	1.0×10^{-6}	cubic meters
cubic centimeters	1.308×10^{-6}	cubic yards
cubic centimeters	2.642×10^{-4}	gallons (U.S. liquid)
cubic centimeters	.0010	liters
cubic centimeters	.002113	pints (U.S. liquid)
cubic centimeters	.001057	quarts (U.S. liquid)
cubic feet	.8036	bushels (dry)
cubic feet	2.8320×10^{4}	cubic cm
cubic feet	1728	cubic inches
cubic feet	.02832	cubic meters
cubic feet	.03704	cubic yards
cubic feet	7.48052	gallons (U.S. liquid)
cubic feet	28.32	liters
cubic feet	59.84	pints (U.S. liquid)
cubic feet	29.92	quarts (U.S. liquid)
cubic feet/min	472.	cubic cm/sec
cubic feet/min	1.699	cubic m/hr
cubic feet/min	.1247	gal/sec

TO CONVERT	MULTIPLY BY	TO OBTAIN
cubic feet/min	.4720	liters/sec
cubic feet/min	62.43	lb water/min
cubic feet/lb	0.06243	m^3/kg
cubic feet/lb	62.43	cm^2/g
cubic feet/sec	.646317	million gal/day
cubic feet/sec	448.831	gal/min
cubic inches	16.39	cubic cm
cubic inches	5.787×10^{-4}	cubic ft
cubic inches	1.639×10^{-5}	cubic meters
cubic inches	5.787×10^{-4}	cubic ft
cubic inches	1.639×10^{-5}	cubic meters
cubic inches	2.143×10^{-5}	cubic yards
cubic inches	.004329	gallons
cubic inches	.01639	liters
cubic inches	.03463	pints (U.S. liquid)
cubic inches	.01732	quarts (U.S. liquid)
cubic meters	28.38	bushels (dry)
cubic meters	1.0×10^{6}	cubic cm
cubic meters	35.31	cubic ft
cubic meters	6.1023×10^{4}	cubic inches
cubic meters	1.308	cubic yards
cubic meters	264.2	gallons (U.S. liquid)
cubic meters	1000	liters
cubic meters	2113.	pints (U.S. liquid)
cubic meters	1057.	quarts (U.S. liquid)
cubic meters/hr	0.5886	cu ft/min
cubic meters/hr at 0°C	0.6223	std cu ft/min at 60°F
cubic meters/hr at 15°C	0.5896	std cu ft/min at 60°F
cubic meters/kilogram	16.02	cu ft/pound
cubic yards	7.646×10^{5}	cubic cm
cubic yards	27.	cubic ft
cubic yards	4.6656×10^{4}	cubic inches
cubic yards	.7646	cu meters
cubic yards	202.	gallons (U.S. liquid)
cubic yards	764.6	liters
cubic yards	1615.9	pints (U.S. liquid)
cubic yards	807.9	quarts (U.S. liquid)
cubic yards/min	.45	cu ft/sec
cubic yards/min	3.367	gal/sec
cubic yards/min	12.74	l/sec

D

daltons	1.650×10^{-24}	grams
days	8.64×10^{4}	seconds
days	1440	minutes
days	24.	hours
decigrams	.10	grams
deciliters	.10	liters
decimeters	.10	meters
degrees (angle)	.01111	quadrants
degrees (angle)	.01745	radians
degrees (angle)	3600	seconds

TO CONVERT	MULTIPLY BY	TO OBTAIN
drams (apoth or troy)	.13714	ounces (avdp.)
drams (apoth or troy)	.125	ounces (troy)
drams (U.S. fluid or apoth)	3.6967	cubic cm
drams	1.7718	grams
drams	27.344	grains
drams	.0625	ounces
dynes/cm^2	9.869×10^{-7}	atmospheres
dynes/cm^2	2.953×10^{-5}	in. of mercury (at 0°C)
dynes/cm^2	4.015×10^{-4}	in. of water (at 4°C)
dynes	.001020	grams
dynes	1.0×10^{-7}	joules/cm
dynes	1.0×10^{-5}	joules/cm (newtons)
dynes	1.020×10^{-6}	kilograms
dynes	7.233×10^{-5}	poundals
dynes	2.248×10^{-6}	pounds
dynes/cm^2	1.0×10^{-6}	bars
	E	
ell	114.30	cm
ell	45.	inches
em,pica	.167	inch
em,pica	.4233	cm
erg/sec	1.0	dyne-cm/sec
ergs	9.486×10^{-11}	BTU
ergs	1.0	dyne-centimeters
ergs	7.376×10^{-8}	foot-pounds
ergs	2.389×10^{-8}	gram-calories
ergs	.001020	gram-centimeters
ergs	3.7250×10^{-14}	horsepower-hr
ergs	1.0×10^{-7}	joules
ergs	2.389×10^{-11}	kg-calories
ergs	1.020×10^{-8}	kg-meters
ergs	2.773×10^{-14}	kilowatt-hr
ergs	2.773×10^{-11}	watt-hr
ergs/sec	5.668×10^{-9}	BTU/min
ergs/sec	4.426×10^{-6}	ft lb/min
ergs/sec	7.3756×10^{-8}	ft lb/sec
ergs/sec	1.341×10^{-10}	horsepower
ergs/sec	1.433×10^{-9}	kg-calories/min
ergs/sec	$1.\times10^{-10}$	kilowatts
	F	
farads	$1.\times10^{6}$	microfarads
faraday/sec	9.65×10^{4}	ampere (absolute)
faradays	26.8	ampere-hours
faradays	9.649×10^{4}	coulombs
fathoms	1.8228	meters
fathoms	6.0	feet
feet	30.48	centimeters
feet	3.048×10^{-4}	kilometers
feet	.3048	meters

TO CONVERT	MULTIPLY BY	TO OBTAIN
feet	1.645×10^{-4}	miles (naut.)
feet	1.894×10^{-4}	miles (stat.)
feet	304.8	millimeters
feet	1.2×10^{4}	mils
feet of water	.0295	atmospheres
feet of water	.8826	in. of mercury
feet of water	.03048	kg/cm^2
feet of water	.4335	lb/in^2
feet/sec	30.48	cm/sec
feet/sec	1.097	km/hr
feet/sec	.5921	knots
feet/sec	.6818	miles/hr
feet/sec/sec	30.48	cm/sec/sec
feet/sec/sec	1.097	km/hr/sec
feet/sec/sec	.3048	meters/sec/sec
feet/sec/sec	.6818	miles/hr/sec
foot-pounds	.00128508	BTU
foot-pounds	.3241	gram-calories
foot-pounds	5.0505×10^{-7}	horsepower-hr
foot-pounds	1.35582	joules
foot-pounds	3.241×10^{-4}	kg-calories
foot-pounds	.1383	kg-meters
foot-pounds	3.766×10^{-7}	kilowatt-hr
foot-pounds/min	.001286	BTU/min
foot-pounds/min	.01667	ft-lb/sec
foot-pounds/min	3.030×10^{-5}	horsepower
foot-pounds/min	3.241×10^{-4}	kg-calories/min
foot-pounds/min	2.260×10^{-5}	kilowatts
foot-$pound_f$/$pound_m$	2.989	newton-meters/kg
foot-$pound_f$/$pound_m$	.002989	kilonewton-m/kg
foot-pounds/sec	4.6263	BTU/hr
foot-pounds/sec	.07717	BTU/min
foot-pounds/sec	.001818	horsepower
foot-pounds/sec	.01945	kg-calories/min
foot-pounds/sec	.001356	kilowatts
furlongs	.125	miles (U.S.)
furlongs	40.	rods
furlongs	660	feet
furlongs	201.17	meters

G

TO CONVERT	MULTIPLY BY	TO OBTAIN
gallons	3785.	cubic cm
gallons	.1337	cubic feet
gallons	231.	cubic inches
gallons	.003785	cubic meters
gallons	.004951	cubic yards
gallons	3.785	liters
gallons (liq. Br. imp.)	1.20095	gallons (U.S. liquid)
gallons (U.S.)	.83267	gallons (imp.)
gallons of water	8.337	pounds of water
gallons/min	.002228	ft^3/sec

TO CONVERT	MULTIPLY BY	TO OBTAIN
gallons/min	.06308	liters/sec
gallons/min	8.0208	ft^3/hr
grams	980.7	dynes
grams	15.43	grains (troy)
grams	9.807×10^{-5}	joules/cm
grams	.009807	joules/meter (newtons)
grams	.0010	kilograms
grams	1000	milligrams
grams	.03527	ounces (avdp.)
grams	.03515	ounces (troy)
grams	.07093	poundals
grams	.002205	pounds
grams/cm	.0056	lb/in
grams/cm^3	.03613	lb/in^3
grams/liter	.062427	lb/ft^3
gram-calories	.0039683	BTU
gram-calories	4.184×10^{7}	ergs
gram-calories	3.086	foot-pounds
gram-calories	1.5596×10^{-6}	horsepower-hr
gram-calories	1.162×10^{-6}	kilowatt-hr
gram-calories	.001162	watt-hr
gram-calories/sec	14.286	BTU/hr
gram-centimeters	9.297×10^{-8}	BTU
gram-centimeters	980.7	ergs
gram-centimeters	9.807×10^{-5}	joules
gram-centimeters	2.343×10^{-8}	kg-calories
	H	
hand	10.16	cm
hectares	2.471	acres
hectares	1.076×10^{5}	square feet
hectograms	100	grams
hectoliters	100	liters
hectometers	100	meters
hectowatts	100	watts
henries	1000	millihenries
hogsheads (British)	10.114	cubic ft
hogsheads (U.S.)	8.42184	cubic ft
hogsheads (U.S.)	63.	gallons (U.S.)
horsepower	42.44	BTU/min
horsepower	3.3×10^{4}	foot-lb/min
horsepower	550.	foot-lb/sec
horsepower (metric)	.9863	horsepower
horsepower	1.014	horsepower (metric)
horsepower	10.68	kg-calories/min
horsepower	.7457	kilowatts
horsepower	745.7	watts
horsepower (boiler)	9.803	kilowatts
horsepower-hours	2544.47	BTU
horsepower-hours	2.6845×10^{13}	ergs
horsepower-hours	1.98×10^{6}	foot-lbs

TO CONVERT	MULTIPLY BY	TO OBTAIN
horsepower-hours	6.4119×10^{5}	gram-calories
horsepower-hours	2.684×10^{6}	joules
horsepower-hours	641.7	kg-calories
horsepower-hours	2.737×10^{5}	kg-meters
horsepower-hours	.7457	kilowatt-hr
hours	.04167	days
hours	.005952	weeks
hours	3600	seconds
hundredwgts(long)	112.	pounds
hundredwgts(long)	.050	tons (long)
hundredwgts(long)	50.8023	kilograms
hundredwgts(short)	.045359	tons (metric)
hundredwgts(short)	.0446429	tons (long)
hundredwgts(short)	45.3592	kilograms

I

TO CONVERT	MULTIPLY BY	TO OBTAIN
inches	2.540	centimeters
inches	.02540	meters
inches	1.578×10^{-5}	miles
inches	25.4	millimeters
inches	1000	mils
inches	.02778	yards
inches	2.54×10^{8}	angstrom units
inches	.0050505	rods
inches of mercury	.03342	atmospheres
inches of mercury	1.1340	feet of water
inches of mercury	.0345316	kg/cm^2
inches of mercury	345.316	kg/m^2
inches of mercury	70.73	lb/ft^2
inches of mercury	.491154	lb/in^2
inches of mercury	.0338639	bar
in. of water (at 4°C)	.002458	atmospheres
in. of water (at 4°C)	.07355	inches of mercury
in. of water (at 4°C)	.00254	kg/cm^2
in. of water (at 4°C)	.5781	$ounces/in^2$
in. of water (at 4°C)	5.204	lb/ft^2
in. of water (at 4°C)	.03613	lb/in^2

J

TO CONVERT	MULTIPLY BY	TO OBTAIN
joules	9.468×10^{-4}	BTU
joules	1.0×10^{7}	ergs
joules	.7736	foot-pounds
joules	2.389×10^{-4}	kg-calories
joules	.1020	kg-meters
joules	2.778×10^{-4}	watt-hr
joules/cm	1.020×10^{4}	grams
joules/cm	1.0×10^{7}	dynes
joules/cm	100	joules/meter (newtons)
joules/cm	723.3	poundals
joules/cm	22.48	pounds

TO CONVERT	MULTIPLY BY	TO OBTAIN
	K	
kilograms	9.80665×10^5	dynes
kilograms	1000	grams
kilograms	.09807	joules/cm
kilograms	9.807	joules/meter (newtons)
kilograms	70.93	poundals
kilograms	2.2046	pounds
kilograms	9.842×10^{-4}	tons (long)
kilograms	.001102	tons (short)
kilograms	35.274	ounces (avdp.)
kg/m^3	.0010	$grams/cm^3$
kg/m^3	.06243	lb/ft^3
kg/m^3	3.613×10^{-5}	lb/in^3
kg/m	.672	lb/ft
kg/cm^2	9.80665×10^5	$dynes/cm^2$
kg/cm^2	.967841	atmospheres
kg/cm^2	32.841	feet of water
kg/cm^2	28.9590	inches of mercury at 32°F
kg/cm^2	2048.	lb/ft^2
kg/cm^2	14.2233	lb/in^2
kg/cm^2	.9807	bars
kg/m^2	9.678×10^{-5}	atmospheres
kg/m^2	9.807×10^{-5}	bars
kg/m^2	.003281	feet of water
kg/m^2	.002896	inches of mercury
kg/m^2	.2048	lb/ft^2
kg/m^2	.001422	lb/in^2
kg/m^2	98.0665	$dynes/cm^2$
kg/mm^2	1.0×10^6	kg/m^2
kilogram-calories	3.968	BTU
kilogram-calories	3086.	foot-pounds
kilogram-calories	.001558	horsepower-hr
kilogram-calories	4183.	joules
kilogram-calories	426.9	kg-meters
kilogram-calories	4.186	kilojoules
kilogram-calories	.001163	kilowatt-hr
kilogram-calories/min	51.43	ft-lbs/sec
kilogram-calories/min	.09351	horsepower
kilogram-calories/min	.06972	kilowatts
kilograms/hour	.03674	lb/min
kilograms/hour	2.205	lb/hr
kilogram-meters	.009295	BTU
kilogram-meters	9.807×10^7	ergs
kilogram-meters	7.23301	foot-pounds
kilogram-meters	9.80665	joules
kilogram-meters	.002342	kg-calories
kilogram-meters	2.723×10^{-6}	kilowatt-hr
kilojoule/kilogram	.4299	BTU/lb_m
kilolines	1000	maxwells
kiloliters	1000	liters
kiloliters	1.308	cubic yards
kiloliters	26.316	cubic feet

TO CONVERT	MULTIPLY BY	TO OBTAIN
kiloliters	264.18	gallons (U.S. liquids)
kilometers	1.0×10^5	centimeters
kilometers	3281.	feet
kilometers	3.937×10^4	inches
kilometers	1000	meters
kilometers	.6214	miles (statute)
kilometers	.5396	miles (nautical)
kilometers	1.0×10^6	millimeters
kilometers	1093.6	yards
kilometers/hr	27.78	cm/sec
kilometers/hr	54.68	ft/min
kilometers/hr	.9113	ft/sec
kilometers/hr	.5396	knots
kilometers/hr	16.67	meters/min
kilometers/hr	.6214	miles/hr
kilometers/hr/sec	27.78	cm/sec/sec
kilometers/hr/sec	.9113	ft/sec/sec
kilometers/hr/sec	.2778	meters/sec/sec
kilometers/hr/sec	.6214	miles/hr/sec
kilonewton-meters/kilogram	334.6	ft-lb_f/lb_m
kilopascals	.145	lb/in^2
kilowatts	56.92	BTU/min
kilowatts	4.426×10^4	foot-lb/min
kilowatts	737.6	foot-lb/sec
kilowatts	1.341	horsepower
kilowatts	14.34	kg-calories/min
kilowatts	1000	watts
kilowatt-hr	3413.	BTU
kilowatt-hr	3.6×10^{13}	ergs
kilowatt-hr	2.655×10^6	foot-lb
kilowatt-hr	8598.5	gram calories
kilowatt-hr	1.341	horsepower-hours
kilowatt-hr	3.6×10^6	joules
kilowatt-hr	860.5	kg-calories
kilowatt-hr	3.671×10^5	kg-meters
kilowatt-hr	3.53	pounds of water evaporated from and at 212°F
kilowatt-hr	22.75	pounds of water raised from 62° to 212°F
knots	6080.	feet/hr
knots	1.8532	kilometers/hr
knots	1.0	nautical miles/hr
knots	1.151	statute miles/hr
knots	2027.	yards/hr
knots	1.689	feet/sec
knots	51.48	cm/sec

L

lambert	.3183	$candle/cm^2$
lambert	2.054	$candle/in^2$
league	3.0	miles(approx.)

TO CONVERT	MULTIPLY BY	TO OBTAIN
light year	5.9×10^{12}	miles
light year	9.46091×10^{12}	kilometers
liters	.02838	bushels (U.S. dry)
liters	1000	cubic cm
liters	.03531	cubic ft
liters	61.02	cubic inches
liters	.0010	cubic meters
liters	.001308	cubic yards
liters	.2642	gallons (U.S. liquid)
liters	2.113	pints (U.S. liquid)
liters	1.057	quarts (U.S. liquid)
liters/min	5.886×10^{-4}	ft^3/sec
liters/min	.004403	gal/sec
$\log_{10}n$	2.303	ln n
ln n	.4343	$\log_{10}n$
lumen	.07958	spherical candle power
lumen/ft^2	1.0	foot-candles
lumen/ft^2	10.76	lumen-m^2
lux	.0929	foot-candles

M

TO CONVERT	MULTIPLY BY	TO OBTAIN
maxwells	.0010	kilolines
maxwells	1.0×10^{-8}	webers
mega pascal, MPa	145.	pounds/in^2
mega watt	1341.0	horsepower
meters	1.0×10^{10}	angstrom units
meters	100	centimeters
meters	.54681	fathoms
meters	3.281	feet
meters	39.37	inches
meters	.0010	kilometers
meters	5.396×10^{-4}	miles (nautical)
meters	6.214×10^{-4}	miles (statute)
meters	1000	millimeters
meters	1.094	yards
meters/min	1.667	cm/sec
meters/min	3.281	feet/min
meters/min	.05468	feet/sec
meters/min	.060	km/hr
meters/min	.03238	knots
meters/min	.03728	miles/hr
meters/sec	196.8	feet/min
meters/sec	3.281	feet/sec
meters/sec	3.6	kilometers/hr
meters/sec	.060	kilometers/min
meters/sec	2.237	miles/hr
meters/sec	.03728	miles/min
meters/sec/sec	100	cm/sec/sec
meters/sec/sec	3.281	ft/sec/sec
meters/sec/sec	3.6	km/hr/sec

TO CONVERT	MULTIPLY BY	TO OBTAIN
meters/sec/sec	2.237	miles/hr/sec
meter-kilograms	9.807×10^{7}	cm-dynes
meter-kilograms	1.0×10^{5}	cm/gram
meter-kilograms	7.233	foot-pound
micromicrons	1.0×10^{-12}	meters
microns	1.0×10^{-6}	meters
miles (nautical)	6076.	feet
miles (nautical)	1.853	kilometers
miles (nautical)	1853.	meters
miles (nautical)	1.1516	miles (statute)
miles (nautical)	2025.4	yards
miles (statute)	1.609×10^{5}	centimeters
miles (statute)	5280.	feet
miles (statute)	6.336×10^{4}	inches
miles (statute)	1.609	kilometers
miles (statute)	1609.	meters
miles (statute)	.8684	miles (nautical)
miles (statute)	1760.	yards
miles (statute)	1.69×10^{-13}	light years
miles/hr	44.70	cm/sec
miles/hr	88.	ft/min
miles/hr	1.467	ft/sec
miles/hr	1.6093	km/hr
miles/hr	.02682	km/min
miles/hr	.8684	knots
miles/hr	26.82	meters/min
milligrams	.0010	grams
milliliters	.0010	liters
millimeters	.10	centimeters
millimeters	.003281	feet
millimeters	.03937	inches
millimeters	1.0×10^{-6}	kilometers
millimeters	.0010	meters
millimeters	6.214×10^{-7}	miles
millimeters	39.37	mils
millimeters	.001094	yards
million gal/day	1.54723	ft^3/sec
mils	.00254	centimeters
mils	8.333×10^{-5}	feet
mils	.0010	inches
mils	2.54×10^{-8}	kilometers
mils	2.778×10^{-5}	yards

N

TO CONVERT	MULTIPLY BY	TO OBTAIN
newtons	1.0×10^{5}	dynes
newtons	.2248	pounds
newton-meter/kilogram	.3346	$ft\text{-}lb_f/lb_m$
newtons/m^3	1.45×10^{-4}	lb/in^2
newtons/m^3	1.0×10^{-5}	bars

TO CONVERT	MULTIPLY BY	TO OBTAIN
	O	
ohms	1.0×1.0^6	microhms
ounces	8.0	drams
ounces	437.5	grains
ounces	28.349	grams
ounces	.0625	pounds
ounces	.9115	ounces/(troy)
ounces	2.790×10^{-5}	tons (long)
ounces	3.125×10^{-5}	tons (short)
ounces (fluid)	1.805	cubic inches
ounces (fluid)	.02957	liters
ounces (troy)	480.	grains
ounces (troy)	31.103	grams
ounces (troy)	1.097	ounces (avdp.)
ounces (troy)	20.	pennyweights (troy)
ounces (troy)	.08333	pounds (troy)
ounce/in^2	4309.	dynes/cm^2
ounce/in^2	.0625	pounds/in^2
	P	
parts/million	.0584	grains/U.S. gal
parts/million	.07016	grains/imp. gal
parts/million	8.345	pounds/million gal
pascals, N/m^3	1.45×10^{-4}	lb/in^2
pascals, N/m^3	10^{-5}	bars
pecks (British)	554.6	cubic inches
pecks (British)	9.0919	liters
pecks (U.S.)	.25	bushels
pecks (U.S.)	537.6	cubic inches
pecks (U.S.)	8.8096	liters
pecks (U.S.)	8	quarts/dry
pennyweights (troy)	24	grains
pennyweights (troy)	.050	ounces (troy)
pennyweights (troy)	1.555	grams
pennyweights (troy)	.0041667	pounds (troy)
pints (dry)	33.6	cubic inches
pints (dry)	.015625	cubic inches
pints (dry)	.50	quarts
pints (dry)	.55059	liters
pints (liquid)	473.2	cubic cm
pints (liquid)	.01671	cubic ft
pints (liquid)	28.87	cubic inches
pints (liquid)	4.732×10^{-4}	cubic meters
pints (liquid)	6.189×10^{-4}	cubic yards
pints (liquid)	.125	gallons
pints (liquid)	.4732	liters
pints (liquid)	.50	quarts (liquid)
Planck's quantum	6.624×10^{-27}	erg-seconds
poise	1.0	gram/cm-sec

TO CONVERT	MULTIPLY BY	TO OBTAIN
pounds (avdp.)	14.583	ounces (troy)
poundals	1.3826×10^{4}	dynes
poundals	14.1	grams
poundals	.001383	joules/cm
poundals	.1383	joules/meter (newtons)
poundals	.0141	kilograms
poundals	.03108	pounds
pounds	256.	drams
pounds	4.448×10^{5}	dynes
pounds	7000	grains
pounds	453.59	grams
pounds	.04448	joules/cm
pounds	4.448	joules/meter (newtons)
pounds	.4536	kilograms
pounds	16.	ounces
pounds	14.58	ounces (troy)
pounds	32.17	poundals
pounds	1.21528	pounds (troy)
pounds	5.0×10^{-4}	tons (short)
pounds (troy)	5760.	grains
pounds (troy)	373.24	grams
pounds (troy)	13.166	ounces (avdp.)
pounds (troy)	12.	ounces (troy)
pounds (troy)	240	pennyweights (troy)
pounds (troy)	.82286	pounds (avdp.)
pounds (troy)	3.6735×10^{-4}	tons (long)
pounds (troy)	3.7324×10^{-4}	tons (metric)
pounds (troy)	4.1143×10^{-4}	tons (short)
pounds of water	.01602	cubic ft
pounds of water	27.68	cubic inches
pounds of water	.1198	gallons
pounds of water/min	2.670×10^{-4}	ft^3/sec
pound-feet	1.356×10^{7}	cm-dynes
pound-feet	1.3825×10^{4}	cm-grams
pound-feet	.1383	meter-kg
pounds/ft^3	.01602	grams/cm^3
pounds/ft^3	16.02	kg/m^3
pounds/ft^3	5.787×10^{-4}	lb/in^3
pounds/ft^3	5.456×10^{-9}	lb/mil-foot
pounds/in^3	27.68	grams/cm^3
pounds/in^3	2.768×10^{4}	kg/m^3
pounds/in^3	1728.	lb/ft^3
pounds/in^3	9.425×10^{-6}	lb/mil-foot
pounds/ft	1.488	kg/meter
pounds/in	178.6	grams/cm
pounds/min	27.22	kilograms/hr
pounds/hr	.4536	kilograms/hr
pounds/mil-foot	2.306×10^{6}	grams/cm^3
pounds/ft^2	4.725×10^{-4}	atmospheres
pounds/ft^2	.01602	feet of water
pounds/ft^2	.01414	inches of mercury
pounds/ft^2	4.882	kg/$meter^2$

TO CONVERT	MULTIPLY BY	TO OBTAIN
pounds/ft^2	.006944	pounds/inch^2
pounds/in^2	.0680460	atmospheres
pounds/in^2	.0689476	bar
pounds/in^2	2.3089	feet of water
pounds/in^2	2.03602	inches of mercury at 32°F
pounds/in^2	703.1	kg/m^2
pounds/in^2	144.	lb/ft^2
pounds/in^2	.0703070	kg/cm^2
pounds/in^2	.006895	mega pascal (MPa)
pounds/in^2	6.895	kilo pascal (kPa), kN/m^2
	Q	
quadrants (angle)	90.	degrees
quadrants (angle)	5400	minutes
quadrants (angle)	1.571	radians
quadrants (angle)	3.24×10^5	seconds
quarts (dry)	67.2	cubic inches
quarts (liquid)	946.4	cubic cm
quarts (liquid)	.03342	cubic ft
quarts (liquid)	57.75	cubic inches
quarts (liquid)	9.464×10^{-4}	cubic meters
quarts (liquid)	.001238	cubic yards
quarts (liquid)	.25	gallons
quarts (liquid)	.9463	liters
	R	
radians	57.296	degrees
radians	3438.	minutes
radians	.6366	quadrants
radians	2.063×10^5	seconds
radians/sec	57.296	degrees/sec
radians/sec	9.549	revolutions/min
radians/sec	.1592	revolutions/sec
radians/sec/sec	572.96	revs/min/min
radians/sec/sec	9.549	revs/min/sec
radians/sec/sec	.1592	revs/sec/sec
reams	500	sheets
revolutions	360.	degrees
revolutions	4.0	quadrants
revolutions	6.283	radians
revolutions/min	6.0	degrees/sec
revolutions/min	.1047	radians/sec
revolutions/min	.01667	revs/sec
rods	.25	chains (gunters)
rods	5.029	meters
rods (surveyors' meas.)	5.5	yards
rods	16.5	feet
rods	198.	inches
rods	.003125	miles
rope	20.	feet

TO CONVERT	MULTIPLY BY	TO OBTAIN
	S	
slugs	14.59	kilograms
slugs	32.17	pounds
square centimeters	.001076	square feet
square centimeters	.1550	square inches
square centimeters	1.0×10^{-4}	square meters
square centimeters	3.861×10^{-11}	square miles
square centimeters	100	square millimeters
square centimeters	1.196×10^{-4}	square yards
square degrees	3.0462×10^{-4}	steradians
square feet	2.296×10^{-5}	acres
square feet	1.883×10^{8}	circular mils
square feet	929.	square cm
square feet	144.	square inches
square feet	.0929	square meters
square feet	3.587×10^{-8}	square miles
square feet	9.29×10^{4}	square millimeters
square feet	.1111	square yards
square inches	1.273×10^{6}	circular mils
square inches	6.452	square cm
square inches	.006944	square ft
square inches	645.2	square millimeters
square inches	1.0×10^{6}	square mils
square inches	7.716×10^{-4}	square yards
square kilometers	247.1	acres
square kilometers	1.0×10^{10}	square cm
square kilometers	1.076×10^{7}	square ft
square kilometers	1.550×10^{9}	square inches
square kilometers	1.0×10^{6}	square meters
square kilometers	.3861	square miles
square kilometers	1.196×10^{6}	square yards
square meters	2.471×10^{-4}	acres
square meters	1.0×10^{4}	square cm
square meters	10.76	square ft
square meters	1550	square inches
square meters	3.861×10^{-7}	square miles
square meters	1.0×10^{6}	square millimeters
square meters	1.196	square yards
square miles	640.	acres
square miles	2.788×10^{7}	square ft
square miles	2.590	square km
square miles	2.590×10^{6}	square meters
square miles	3.098×10^{6}	square yards
standard ft^3/min at 60°F	1.607	m^3/hr at 0°C
standard ft^3/min at 60°F	1.696	m^3/hr at 15°C
square millimeters	1973.	circular mils
square millimeters	.010	square cm
square millimeters	1.076×10^{-5}	square ft
square millimeters	.00155	square inches
square mils	1.273	circular mils
square mils	6.452×10^{-6}	square cm

TO CONVERT	MULTIPLY BY	TO OBTAIN
square mils	1.0×10^{-6}	square inches
square yards	2.066×10^{-4}	acres
square yards	8361.	square cm
square yards	9.0	square ft
square yards	1296.	square inches
square yards	.8361	square meters
square yards	3.228×10^{-7}	square miles
square yards	8.361×10^{5}	square millimeters
	T	
temperature (°C) + 273	1.0	absolute temperature (°K)
temperature (°C) + 17.78	1.8	temperature (°F)
temperature (°F) + 460	1.0	absolute temperature (°R)
temperature (°F) − 32	5/9	temperature (°C)
tonne	1000	kilograms
tonne	2025	pounds
tonne	1.0	(metric) tons
tons (long)	1016	kilograms
tons (long)	2240	pounds
tons (long)	1.12	tons (short)
tons (metric)	1000	kilograms
tons (metric)	2205	pounds
tons (short)	907.18	kilograms
tons (short)	3.2×10^{4}	ounces
tons (short)	2.9166×10^{4}	ounces (troy)
tons (short)	2000	pounds
tons (short)	2430	pounds (troy)
tons (short)	.89287	tons (long)
tons (short)	.9078	tons (metric)
tons (short)/ft^2	9765	kg/m^2
tons (short)/ft^2	13.89	lb/in^2
tons (short)/in^2	1.406×10^{6}	kg/m^2
tons (short)/in^2	2000	lb/in^2
	W	
watts	3.4129	BTU/hr
watts	.05688	BTU/min
watts	1.0×10^{7}	ergs/sec
watts	44.27	ft-lb/min
watts	.7378	ft-lb/sec
watts	.001341	horsepower
watts	.00136	horsepower (metric)
watts	.01433	kg-calories/min
watts	.0010	kilowatts
watts (abs.)	1.0	joules/sec
watt-hours	3.413	BTU
watt-hours	3.6×10^{10}	ergs
watt-hours	2656.	foot-lb
watt-hours	860.5	gram-calories
watt-hours	.001341	horsepower-hours

TO CONVERT	MULTIPLY BY	TO OBTAIN
watt-hours	.8605	kilogram-calories
watt-hours	367.2	kilogram-meters
watt-hours	.0010	kilowatt-hours
webers	1.0×10^8	maxwells
webers	1.0×10^5	kilolines
webers/in^2	1.55×10^7	gausses
webers/in^2	1.0×10^8	lines/in^2
webers/in^2	.155	webers/cm^2
webers/in^2	1550	webers/m^2
webers/m^2	1.0×10^4	gausses
webers/m^2	6.452×10^4	lines/in^2
webers/m^2	1.0×1.0^{-4}	webers/cm^2
webers/m^2	6.452×10^{-4}	webers/in^2
weeks	168.	hours
weeks	1.008×10^4	minutes
weeks	6.048×10^5	seconds
	Y	
yards	91.44	centimeters
yards	9.144×10^{-4}	kilometers
yards	.9144	meters
yards	4.934×10^{-4}	miles (nautical)
yards	5.682×10^{-4}	miles (statute)
yards	914.4	millimeters
years	365.256	days (mean solar)
years	8766.1	hours (mean solar)

ENGINEERING CONSTANTS

1 HP = 33,000 ft-lb/min

1 HP = .7457 K.W.
= K.W./1.341

1 HP = 2546.4 BTU per hour

1 BTU = Heat required to raise 1 lb. water 1°F

1 BTU = 778.16 Foot-pounds

1 kilowatt hour = 3413 BTU

Heat value of carbon = 14,600 BTU per pound

Latent heat of fusion of ice = 143.15 BTU per pound

Latent heat of evaporation of water at 212°F = 970.4 BTU per pound

Total heat of saturated steam at atmospheric pressure = 1,150.4 BTU per pound

1 ton of refrigeration = 288,000 BTU per 24 hours

Gas constant (R)

$$R = 1545.32 \frac{\text{ft-lb}}{\text{Lb-Mole-}^\circ\text{R}}$$

$$R = 8.314 \frac{\text{joules}}{\text{gm-Mole-}^\circ\text{K}}$$

1 radian = 57.296 degrees

1 meter = 100 cm = 39.37 inches

1 kilometer = .62137 miles

1 gallon = 231 cubic inches

1 barrel = 31.5 gallons

Atmospheric pressure
= 14.7 pounds per square inch
= 29.92 inches mercury at 32°F

1 pound per square inch pressure
= 2.3095 feet fresh water at 62°F
= 2.0355 inches mercury at 32°F
= 2.0416 inches mercury at 62°F

Water pressure (pounds per square inch) = .433 × height of water in feet (fresh water at 62°F)

Weight of 1 cubic foot fresh water = 62.355 lb at 62°F = 59.76 lb at 212°F

Weight of 1 cubic foot air at 14.7 lb per square inch pressure = .07608 lb at 62°F = .08703 lb at 32°F

Velocity of sound in dry air at 0°C and 1 atm = 33,136 cm/sec = 1089 ft/sec

Acceleration of gravity (standard)
= 32.17 ft/sec^2
= 980.6 cm/sec^2

References

1. Bodger WK, Jensen RC, *Fundamentals of Fluid Flow as Applied to the Design of Axial Flow Compressors and Fans,* Carrier Corp., Syracuse, NY, and Elliott Co., Jeannette, PA, 1954.
2. Wilson DG, *The Design of High-Efficiency Turbomachinery and Gas Turbines,* MIT Press, Cambridge, MA, 1984.
3. Sawyer RT, *The Modern Gas Turbine,* Prentice-Hall, New York, 1945.
4. Adams HT, *Elements of Internal Combustion Turbine Theory,* Canterbury University, New Zealand, 1949.
5. Hill PG (MIT), Peterson CR (Ingersoll-Rand Co.), *Mechanics and Thermodynamics of Propulsion,* Addison-Wesley, Reading, MA, 1965.
6. Vietmeyer N, *They Created the Jet Age,* Reader's Digest, Pleasantville, NY, May 1987.
7. *Centrifugal Compressors for General Refinery Service, 5th Ed.,* API Standard 617, American Petroleum Institute, Washington D.C., 1988.
8. Baumeister T, *Standard Handbook for Mechanical Engineers,* McGraw-Hill, New York, 1967.
9. Sheperd DG (Cornell University), *Principles of Turbomachinery,* Macmillan, New York, 1956.
10. Paluselli DA, *Basic Aerodynamics of Centrifugal Compressors,* Elliott Co., Jeannette, PA.
11. *Compressor Refresher,* Elliott Co., Jeannette, PA, 1975.
12. Shames IH (University of Buffalo), *Mechanics of Fluids,* 1962.
13. Van Wylen G, Sonntag R (University of Michigan), *Fundamentals of Classical Thermodynamics,* John Wiley and Sons, New York, 1968.
14. Derrickson GW, *Thermodynamic Review,* Elliott Co., Jeannette, PA, 1968.
15. *The Orifice Meter,* Rockwell Manufacturing Co., Pittsburgh, PA, 1938.
16. Sassos M, *Compressor Components,* Elliott Co., Jeannette, PA, 1986.
17. Hallock DC, *Centrifugal Compressors . . . The Cause of the Curve,* Elliott Co., Jeannette, PA, 1968.
18. Salisbury R, *Compressor Performance,* Elliott Co., Jeannette, PA, 1986.
19. Whiteman P, *Axial Compressor Design Philosophy,* Elliott Co., Jeannette, PA.
20. Lapina R, *Can You Rerate Your Centrifugal Compressor?* Elliott Co., Jeannette, PA, 1975.
21. *Power and Test Codes, Compressors and Exhausters,* ASME PTC 10, American Society of Mechanical Engineers, New York, 1965.
22. Dunaway J, *Guidelines for Mechanical Field Testing of Compressors,* Elliott Co., Jeannette, PA, 1979.
23. Hackel R, King R, *Centrifugal Compressor Inlet Piping—A Practical Guide,* Elliott Co., Jeannette, PA, 1977.
24. Bensema D, *Field Performance Testing,* Elliott Co., Jeannette, PA, 1986.
25. Leipman HW, Roshko A (California Institute of Technology), *Elements of Gas Dynamics,* John Wiley and Sons, New York, 1957.
26. Lock JA, *Techniques for More Accurate Centrifugal Compressor Performance Evaluation,* Southwest Research, Turbomachinery Symposium, Houston, TX, 1981.
27. Ishida M (Nagasaki University), Senoo Y (Kyushu University), *The Pressure Losses due to the Tip Clearance of Centrifugal Blowers,* American Society of Mechanical Engineers, New York, 1980.
28. Lakshminarayana B (Pennsylvania State University), *Methods of Predicting the Tip Clearance Effects in Axial Flow Turbomachinery,* Journal of Basic Engineering, American Society of Mechanical Engineers, New York, 1970.
29. *Fluid Meters,* ASME PTC 19.5, American Society of Mechanical Engineers, New York, 1971.
30. *Elliott Multistage Centrifugal Compressors,* Bulletin P-25C, Elliott Co., Jeannette, PA, 1985.
31. *Engineering Data Book,* Natural Gas Processors Suppliers Association, Tulsa, OK, 1967, 1972.
32. *Properties of Commonly Used Refrigerants,* Air Conditioning and Refrigeration Institute, Washington D.C., 1957.

Additional Reading

Church AH (New York University), *Centrifugal Pumps and Blowers,* John Wiley and Sons, New York, 1967.

Fergerson TB (University of Sheffield), *The Centrifugal Compressor Stage,* Butterworths, London, 1963.

Flow of Fluids through Valves, Fittings, and Pipe, Crane Technical Paper no. 410, Crane Co., New York, 1978.

Fluid Flow Data Book, General Electric Corporate Research and Development, Schenectady, NY, 1975.

Laboratory Methods of Testing Fans for Rating, AMCA 210-74 and ASHRAE 51–75, American Society of Heating, Refrigerating and Air Conditioning Engineers, Inc., New York, 1974.

Lapina RP, *Estimating Centrifugal Compressor Performance,* Gulf Publishing, Houston, TX, 1982.

Lee BI, Kesler MG, *A Generalized Thermodynamic Correlation Based on Three-Parameter Corresponding States,* Mobil Research and Development Corp., ALChE (May), 1975.

Kunkle J, Wilson S, Cota R (eds.), *Compressed Gas Handbook,* NASA, Washington D.C., 1969.

Nisenfeld AE, *Centrifugal Compressors—Principles of Operation and Control,* Instrument Society of America, Research Triangle Park, NC, 1982.

Pichot P, *Compressor Application Engineering,* Gulf Publishing, Houston, TX, 1986.

Spink LK, *Principles and Practice of Flow Meter Engineering,* Foxboro, MA, 1967.

Starling KE, *Fluid Thermodynamic Properties for Light Petroleum Systems,* Gulf Publishing, Houston, TX, 1973.

Walas SM (University of Kansas), *Phase Equilibria in Chemical Engineering,* Butterworths, Stoneham, MA, 1985.

Compressor Performance Software

Software is available for your IBM-PC compatible computer for performance calculations. This program handles compressor selection as well as performance testing. Some functions include compressor roughout, gas properties, head, flow, and efficiency calculations plus plotting of results. The user will find the program easy to use as it is menu driven and contains appropriate help files.

For order details or for further information on program capabilities, please write to:

Pacific Turbosystems, Inc.
10180 Larrylyn Drive
Whittier, CA 90603

Index